Environmental Expertise

An important goal of environmental research is to inform policy and decision making. However, environmental experts working at the interface between science, policy, and society face complex challenges, including how to identify sources of disagreement over environmental issues, communicate uncertainties and limitations of knowledge, and tackle controversial topics such as genetic modification or the use of biofuels. This book discusses the problems environmental experts encounter in the interaction between knowledge, society, and policy on both a practical and a conceptual level. Key findings from social science research are illustrated with a range of case studies, from fisheries to fracking. The book offers guidance on how to tackle these challenges, equipping readers with tools to better understand the diversity of environmental knowledge and its role in complex environmental issues. Written by leading natural and social scientists, this text provides an essential resource for students, scientists, and professionals working at the science–policy interface.

ESTHER TURNHOUT is a professor at the Forest and Nature Conservation Policy Group of Wageningen University, the Netherlands. Her research programme, 'The Politics of Environmental Knowledge', includes research into the different roles experts play at the science–policy interface, the political implications of policy-relevant environmental knowledge, and the interaction between science, society, and citizens.

WILLEMIJN TUINSTRA works as an independent advisor on knowledge development for environmental policy. She advises research institutes and civil servants on procedural aspects of stakeholder participation, scenario development, and dealing with uncertainties. She has been involved in projects for the European Environment Agency and the International Institute for Applied Systems Analysis.

WILLEM HALFFMAN is an associate professor in the Faculty of Science at Radboud University, Nijmegen. His work focuses on studying how scientific knowledge is validated and how scientists advise public decision making. He has more than 20 years' experience teaching science and society courses to natural scientists.

"The novel perspectives in this book, as well as its cases and conceptual advances, will come as a welcome resource for those interested in understanding the controversies around the use of environmental expertise. Equally, it will be welcomed by environmental scientists seeking to navigate the shoals of practicing and representing science on the one hand, and effective communication and making a difference in the world on the other."

– Arun Agrawal, University of Michigan

"This book is important and timely. There has never before been such great need for evidence to underpin environmental policy, and yet there is also a growing appreciation among researchers of the complexities and risks associated with engaging with the policy community. The book is rooted in the latest theoretical understandings from social science, explained in an accessible way, and it very quickly moves from theory to practice, showing how these insights can inform how environmental scientists work around controversial topics. The use of in-depth case studies complements the widespread use of helpful examples throughout the text. I have found reading this book both inspiring and instructive, and believe that many researchers will benefit considerably from reading it. I will certainly be recommending it to colleagues."

– Mark Reed, Newcastle University

"The principal authors have done a very smart and novel job with this book, bringing together insights from environmental science, policy studies, science studies and the philosophy of science in a thoroughly practical way. The book should really help practitioners appreciate ways to handle the complexities of environmental policy-making in contexts of uncertainty, conflicting beliefs and competing societal values."

– Steve Yearley, IASH, University of Edinburgh

Environmental Expertise
Connecting Science, Policy, and Society

ESTHER TURNHOUT
Wageningen University, the Netherlands
WILLEMIJN TUINSTRA
Open Universiteit, the Netherlands
WILLEM HALFFMAN
Radboud University, Nijmegen, the Netherlands

With contributions from
Silke Beck, Heleen de Coninck, Thomas Gieryn, Mike Hulme, Marga
Jacobs, Phil Macnaghten, Clark Miller, Katja Neves, Martin Pastoors,
Ad Ragas, Claire Waterton, and Laurence Williams

CAMBRIDGE
UNIVERSITY PRESS

CAMBRIDGE
UNIVERSITY PRESS

University Printing House, Cambridge CB2 8BS, United Kingdom

One Liberty Plaza, 20th Floor, New York, NY 10006, USA

477 Williamstown Road, Port Melbourne, VIC 3207, Australia

314-321, 3rd Floor, Plot 3, Splendor Forum, Jasola District Centre, New Delhi - 110025, India

103 Penang Road, #05-06/07, Visioncrest Commercial, Singapore 238467

Cambridge University Press is part of the University of Cambridge.

It furthers the University's mission by disseminating knowledge in the pursuit of education, learning and research at the highest international levels of excellence.

www.cambridge.org
Information on this title: www.cambridge.org/9781107491670
DOI: 10.1017/9781316162514

© Esther Turnhout, Willemijn Tuinstra and Willem Halffman 2019

First published 2019

A catalogue record for this publication is available from the British Library

ISBN 978-1-107-09874-9 Hardback
ISBN 978-1-107-49167-0 Paperback

Additional resources for this publication at www.cambridge.org/turnhout.

Contents

Contributors

Silke Beck
Helmholtz Centre for Environmental Research – UFZ, Leipzig, Germany

Heleen de Coninck
Department of Environmental Science, Radboud University, Nijmegen, the Netherlands

Willem Halffman
Institute for Science in Society, Radboud University, Nijmegen, the Netherlands

Mike Hulme
Department of Geography, University of Cambridge, Cambridge, UK

Thomas Gieryn
Department of Sociology, Indiana University, Bloomington, IN, USA

Marga Jacobs
Faculty of Management, Science and Technology (MST), Open Universiteit, Heerlen, the Netherlands

Phil Macnaghten
Knowledge, Technology and Innovation Group, Wageningen UR, Wageningen, the Netherlands

Clark Miller
School for the Future of Innovation in Society, Arizona State University, Tempe, Arizona, USA

Katja Neves
Department of Sociology and Anthropology, Concordia University, Montreal, Canada

Martin Pastoors
Chief Science Officer, Pelagic Freezer-trawler Association, Zoetermeer, the Netherlands

Ad Ragas
Faculty of Management, Science and Technology (MST), Open Universiteit, Heerlen, the Netherlands

Willemijn Tuinstra
Faculty of Management, Science and Technology (MST), Open Universiteit, Heerlen, the Netherlands

Esther Turnhout
Forest and Nature Conservation Policy Group, Wageningen UR, Wageningen, the Netherlands

Claire Waterton
Department of Sociology, Lancaster University, Lancaster, UK

Laurence Williams
Science Policy Research Unit, University of Sussex, Brighton, UK

Preface

This book was designed to prepare environmental scientists for work in, and with, society. It covers a wide range of examples and case studies, from fisheries and biodiversity conservation to climate change and pesticide pollution, to cover the kinds of issues environmental scientists might work on, as well as providing a healthy dose of conceptual understanding of the problems involved. The primary aim of the book is therefore educational, aiming specifically at environmental scientists with a natural science background, on a Master's or early career researcher level, but also at practitioners in environmental advice or policy.

As such, the book aims to complement the kind of knowledge that dominates natural science education: that is, how to solve well-structured, clearly defined problems with objective and instrumental knowledge, using trusted methods. However, in a societal context, problems are often *ill-structured*, problem definitions are *contested*, facts and methods may be *controversial*, and science is *not automatically trusted* (and often with good reason, because science has created problems in the past, such as organochlorine pesticides, nuclear waste, failed planning megalomania, or plastic soup).

When faced with a critical society, even the most powerful and well-intentioned drive to find optimal solutions for environmental problems and support instrumental reasoning may run into often unforeseen problems and resistance. For example, to many scientists' surprise and disappointment, genetic modification raised fundamental objections, the promises of biofuels were unexpectedly challenged by NGOs, and meticulous assessments of environmental problems were simply pushed aside as irrelevant. This book provides insights that will help us to understand and cope with these dynamics. We need to stress that our aim is not to negate the value of the sciences and their potential for emancipation and the betterment of humanity, but to better accommodate their

xi

particular modes of operation with societal deliberation and collective problem solving.

Many educational programmes in the natural sciences (and the environmental sciences in particular) currently include courses to reflect on the role of scientists in society – courses such as 'Science in Society', 'The Social Responsibility of the Scientist', and 'Environmental Science and Policy'. Using insights from the social sciences, such courses explain how the logic of science-in-society differs from that of science-in-the-lab, and especially from science-in-the-textbook. They prepare scientists to become professionals: academically trained researchers operating in a society that is critical, ever more highly educated, at times also blinded by passion or prejudice, political, mediatised, divided, demanding, concerned. This book is intended to offer reflection and deeper conceptual understanding, as well as practical advice for use in such courses.

We hope the book will also be useful for self-study or reflection by environmental professionals working for research, advisory, policy, industry, or civil society organisations. A primary focus of the book is on the complications of providing science-based advice to collective decision making, either in cooperation with governments or in governance configurations that involve civil society and/or companies. It describes how scientific advice comes to be seen as useful, but also how science gets challenged in environmental controversies.

The book's insights and approach are based on the authors' many years of experience in interdisciplinary teaching and of working in professional environments between science and policy, combining social- and natural-science ways of thinking. All of our contributors have taught courses on environmental controversies or 'science in society' for natural scientists, or have tried to engage in the difficult conversation between social and natural science conceptions of environmental expertise. In fact, in our view, it is especially after some experience of working as a professional that environmental scientists and experts come to experience the limitations of throwing certified facts at society to make it do things. Pointing out the problems *ahead* of such experience, to students trained in laboratories and computer models alone, is both necessary and also much harder.

Our book tries to fill a gap we ourselves experienced when looking for reading material for these difficult courses. We wanted to combine the conceptual work that inspired our own research (and that is often very critical of the way the sciences are currently deployed) with practical advice on how to operate as an environmental science professional, but in ways that are sensitive to the academic criticism and conceptual contributions. Prominent among the conceptual inspiration are critical understandings in interpretative Policy Studies or Science and Technology Studies (such as Flyvbjerg, 1998; Jasanoff, 2005; Scott, 1998; Wynne, 1996). This book complements and differs from Science and Technology Studies handbooks (e.g. Felt, Fouché, Miller, and Smith-Doerr, 2016; Sismondo, 2004; Yearley, 2005) by translating some of these insights specifically for communication with the natural sciences. We are well aware that this may have come at the expense of theoretical sophistication, but that is a risk we are willing to take for the benefit of wider application.

At the practical end of the spectrum, we also made use of more concrete instructions for environmental scientists. These include protocols or checklists for handling uncertainty (Halffman and Ragas, 2016; Petersen et al., 2013; Van Asselt, 2000; Van der Sluijs et al., 2004), for writing expert policy reports, for participating in expert committees, for organising expert advice (Commission of the European Communities, 2002; UK Government, 2005; Young, Watt, and Van den Hove, 2013), and even for how to address the media or position oneself with respect to policy makers (Pielke, 2007). We combined such practical instruction with the theoretical and conceptual rationale behind these instructions, as reflective practitioners should be able to derive their own principles, even in situations that are entirely new and for which there are no protocols (Schön, 1995).

As an educational text, the book is intended as an integrated package that builds on shared conceptual foundations in interpretative social sciences, comprising chapters describing specific issues or providing practical advice. It starts with a conceptual understanding of the nature of scientific and societal framing of problems, and the limitations of the attempt to rationalise the world in an instrumental way (Introduction and Chapter 2 on the nature of the sciences). From experience, we know that some students find these concerns too remote or abstract, in which case it may be useful to return to Chapters 2 and 3 at a later stage. The later chapters in the book focus on more specific issues, such as controversies, the challenges of creating 'integrated' knowledge, and the value of non-scientific forms of lay or local knowledge. Most chapters have at least one case study, providing more elaborate examples, illustrative practical experience, or deeper understanding of the material discussed in the main text. If used as a textbook, lecturers may wish to select the cases closest to their students' interests. We also plan to provide more such material on the book's website and facilitate the sharing of such material among its users. We should point out that the book provides course material, but does not impose a specific instructional method, such as problem-based learning, seminar discussions, or flipped classrooms. Hence, there are no schedules, study questions, or tests provided in the book, although such tools can be shared online. We would be very keen to hear about your experience of using the book and welcome suggestions for improvement.

Writing this book has been a long process, involving the efforts of many authors and reviewers. The idea for this book was born out of a Dutch Open University course, in which we were able to try out the approach and some of the materials. We would like to thank Joop de Kraker and Ron Cörvers of the Open University for the original assignment to develop a course. It had always been Joop's intention to develop a book related to the course, while Ron's original idea can be recognised in the book's approach: starting with problems, to the organisation of knowledge, via the individual role of experts, towards approaches to improved knowledge development. In addition, some of the original work of the Dutch Open University would not have been possible without the contributions of Bertien Broekhans. At the same time, this book takes an extra step from the original course text: all of the material was thoroughly rewritten and supplemented with extra case descriptions and chapters, to accommodate an international audience. We would also like to thank Dave Huitema of the Dutch Open University and Emma Kiddle of Cambridge University Press for convincing us to turn this into a book project. Special thanks also to Zoë Pruce for guiding us through the publication process. We further need to thank our patient friends and colleagues for continuing to enquire about the book's progress and for shaming us into its completion. We thank the Dutch Open University for allowing us to reuse Table 4.1 in Chapter 4 and parts of the text in Chapters 6 and 9, which draw on the original course materials.

We are grateful to the book's reviewers, Alan Irwin, Mark Reed, and Steve Yearley, along with colleagues who reviewed chapters or otherwise helped us along with suggestions and ideas, including Marjolein van Asselt, Rob Maas, Claire Marris, and Andrew Stirling. As editors, we thank the many patient authors who contributed case descriptions and editorial comments on the main text, and Jerry van Dijk for his support in preparing some of the images. Last, Willem Halffman would like to thank the Science, Technology and Innovation Studies unit and the Institute for Advanced Studies in the Humanities at Edinburgh University for their hospitality and stimulating environment during writing breaks in 2015 and 2016.

References

Commission of the European Communities. (2002). *Communication from the Commission on the Collection and Use of Expertise by the Commission: Principles and Guidelines: Improving the Knowledge Base for Better Policies*. '*Improving the Knowledge Base for Better Policies*', COM(2002) 713 final. http://ec.europa.eu/governance/docs/comm_expertise_en.pdf

Felt, U., Fouché, R., Miller, C. A., and Smith-Doerr, L., eds. (2016). *The Handbook of Science and Technology Studies, Fourth Edition*. Cambridge, MA: MIT Press.

Flyvbjerg, B. (1998). *Rationality and Power: Democracy and Practice*. Chicago: University of Chicago Press.

Halffman, W., and Ragas, A. M. (2016). *Achter de Horizon: Omgaan met onzekerheid bij nieuwe risico's*. Den Haag: www.rijksoverheid.nl.

Jasanoff, S. (2005). *Designs on Nature: Science and Democracy in Europe and the United States*. Princeton: Princeton University Press.

Petersen, A. C., Janssen, P. H. M., van der Sluijs, J. P., et al. (2013). *Guidance for Uncertainty Assessment and Communication*. The Hague: PBL.

Pielke, R. (2007). *The Honest Broker: Making Sense of Science in Policy and Politics*. Cambridge: Cambridge University Press.

Schön, D. (1995). *The Reflective Practitioner: How Professionals Think in Action*. London: Arena.

Scott, J. C. (1998). *Seeing like a State: How Certain Schemes to Improve the Human Condition Have Failed*. New Haven: Yale University Press.

Sismondo, S. (2004). *An Introduction to Science and Technology Studies*. Malden: Blackwell.

UK Government. (2005). *Guidelines on Scientific Analysis in Policy Making*. http://webarchive.nationalarchives.gov.uk/20070402091927/http://www.dti.gov.uk/science/science-in-govt/works/advice-policy-making/guidelines/page9474.html

Van Asselt, M. B. A. (2000). *Perspectives on Uncertainty and Risk: The PRIMA Approach to Decision Support*. Dordrecht: Kluwer.

Van der Sluijs, J. P., Janssen, P. H. M., Petersen, A. C., Kloprogge, P., Risbey, J. S., and Tuinstra, W. (2004). *RIVM/MNP Guidance for Uncertainty Assessment and Communication: Tool Catalogue for Uncertainty Assessment*. Utrecht University.

Wynne, B. (1996). May the Sheep Safely Graze? A Reflexive View of the Expert–Lay Knowledge Divide. In S. Lash, B. Szerszynski, and B. Wynne, eds., *Risk, Environment and Modernity: Towards a New Ecology* (pp. 44–83). London: SAGE Publications.

Yearley, S. (2005). *Making Sense of Science: Understanding the Social Study of Science*. London: SAGE Publications.

Young, J. C., Watt, A. D., and Van den Hove, S. (2013). *Effective Interfaces between Science, Policy and Society: The SPIRAL Project Handbook*. www.spiral-project.eu.

Abbreviations

BGCI	Botanical Gardens Conservation International
BECCS	Bio-energy with carbon capture and storage
BEIS	Department for Business, Energy and Industrial Strategy
CBA	Cost–benefit analysis
CBD	Convention on Biological Diversity
CCS	Carbon dioxide capture and storage
CCVS	*Conservatoire des Collections Végétales Spécialisées*
CDM	Clean development mechanism
CERN	European Organization for Nuclear Research
CFCs	Chlorofluorocarbons
CIAM	Conventions' Centre for Integrated Assessment Modelling
CLRTAP	Convention on Long-range Trans-boundary Air Pollution
CUDOS	Communalism, Universalism, Disinterestedness, Organised Scepticism scientific norms
DECC	Department of Energy and Climate Change
DG	Directorate General (of the European Commission)
ECN	Energy Research Centre of the Netherlands
EIA	Environmental impact assessment
EMEP	Cooperative Program for Monitoring and Evaluation of the Long-range Transmission of Air Pollutants in Europe
ETS	Emissions trading scheme
GM	Genetic modification
GMO	Genetically modified organism
GSPC	Global Strategy for Plant Conservation
IAMs	Integrated Assessment Models
IPBES	Intergovernmental Platform for Biodiversity and Ecosystem Services

IPCC	Intergovernmental Panel on Climate Change
LCP	Loweswater Care Project
LEK	Local ecological knowledge
LRTAP	Long-range transboundary air pollution
MA	Millennium Ecosystem Assessment or Millennium Assessment
MCA	Multi-criteria analysis
MNP	*Milieu en NatuurPlanbureau*, Netherlands Environmental Assessment Agency, now PBL
NGO	Non-governmental organisation
OPEC	Organisation for Petroleum Exporting Countries
PBL	*Plan Bureau voor de Leefomgeving*, Netherlands Environmental Assessment Agency
PLACE	Proprietary, local, commissioned, expert counter-norms
RAINS	Regional acidification information and simulation model
R&D	Research and development
RIVM	*Rijksinstituut voor Volksgezondheid en Milieu*, Netherlands National Institute for Public Health and the Environment
SPM	Summary for Policymakers
SRCCS	Special Report on Carbon Dioxide Capture and Storage
STS	Science and Technology Studies
TBG	Toronto Botanical Gardens
TEK	Traditional ecological knowledge
TFIAM	Task Force on Integrated Assessment Modelling
UN	United Nations
UNECE	United Nations Economic Commission for Europe
UNEP	United Nations Environment Programme
UNFCCC	United Nations Framework Convention on Climate Change
VOCs	Volatile organic compounds
WBGU	German Scientific Advisory Council for Global Environmental Change
WHO	World Health Organization
WMO	World Meteorological Organization

1

Introduction

The Plight of the Environmental Scientist

WILLEM HALFFMAN, ESTHER TURNHOUT,
AND WILLEMIJN TUINSTRA

This is a book about how environmental knowledge is used in policy, and how it is transformed to be useful for public problem solving. It is also about how such processes sometimes fail, or are based on misguided conceptions of science, of policy, or of public concerns about environmental matters. We will describe the problems environmental professionals encounter in the interaction between knowledge, policy, and society, on a practical as well as deeper, conceptual levels. Ultimately, this books aims to offer guidance on how experts can play a productive role in the governance of current environmental challenges, while respecting the diversity of perspectives and knowledge claims, and taking into account the concerns of a democratic society.

To do so, the book builds on the knowledge and experience of both social and natural scientists, and tries to combine these insights without reducing them to the lowest common denominator. Rather than providing simple rules of thumb, we want to explain the logic behind them. This chapter explains why and how the book addresses these issues.

1.1 Science and the Environment

Environmental scientists have their work cut out for them, as humanity faces daunting environmental challenges. Climate change is endangering the livelihood of millions, overfishing and plastic pollution are threatening our oceans, while fertile land and biodiversity are under pressure. In turn, environmental problems lead to conflict, scarcity leads to a global scramble for natural resources, and the combined effects of environmental degradation disproportionally impact poor and vulnerable communities. The expectations for environmental sciences to help us understand and solve these problems are high.

To help us comprehend and cope with such challenges, we clearly *need* the sciences. We will also need wisdom, ingenuity, dogged environmental activism, political commitment, solidarity, and probably a bit of luck, but the sciences will remain an essential element in solving environmental problems. Without science, it is impossible to trace pollution that is too small to observe directly, or to establish causal connections between environmental processes that occur on a planetary scale.

One such connection is the link between CFCs (chlorofluorocarbons) and the depletion of the ozone layer. Until the 1980s, CFCs were a main ingredient in spray cans and cooling liquids, and they caused a dangerous degradation of Earth's protective ozone layer, especially over the Antarctic. Only scientific research, with its satellite measurements, its understanding of atmospheric chemistry, and its global research networks, could have established the relation between kitchen-fridge coolant and the atmosphere high above a remote and inhospitable continent. And the world actually listened: 30 years after an international convention, shored up with technological alternatives and regulatory commitment, the ozone layer finally seems to be 'healing'. When nature fails to speak to us unambiguously, we need scientists to provide a translation.

1.2 The Challenges for Environmental Professionals

Yet clearly, the authority of scientists to speak for the environment is not self-evident. Often, environmental scientists express concern or even despair over 'governments who do not listen', or 'the public who do not understand the facts', or even politicians who no longer even seem to *care* about facts. One reason could be that environmental scientists are often the bearers of bad news, and that shooting the messenger may be easier than actually addressing the problem. The case of the L'Aquila earthquake (see example given below) illustrates the difficult position of environmental experts. Luckily, the consequences are rarely as extreme as in this case, but the example does indicate how difficult the position of the expert can become.

The L'Aquila Earthquake: A Trial Against Science?

ESTHER TURNHOUT

The case of the 2009 L'Aquila earthquake in Italy is a useful illustration of some of the thorny issues and dilemmas that experts face. This earthquake not only ended up taking hundreds of lives and destroying historic buildings, it also became important in redefining the relation between expertise and society. In the aftermath of the earthquake, experts were put on trial. The seven members of the national committee for major risks were charged with involuntary manslaughter, on the basis that they had unjustifiably reassured the public and downplayed the risks of the expected earthquake. In 2012, these experts were found guilty and sentenced to six-year prison terms. In 2014, the appeals court reduced the sentence of one of the experts to two years and acquitted the other six. Nevertheless, scientists worldwide were shocked at the thought they could end up in jail for failing to deliver proper advice.

The case triggered a variety of responses in the news media. Most coverage of the case was quick to side with the scientists, comparing their fate with that of Galileo and claiming that it was unreasonable to send scientists to jail for failing to predict the earthquake (Fox News, 2012; Kington, 2012; Davies, 2012). Coverage also focused on the implications of the verdict for the independence and autonomy of science, and on how the verdict might affect the willingness of scientists to sit in policy committees and offer advice (Davies, 2012; Brown, 2012). However, a closer look shows that there was more to this case than simply a matter of the rationality of science being threatened by ignorance and interests. Rather, several lessons can be learnt from the case, with wider significance for the connections between science, policy, and society.

First, the court case was not about scientific predictions and calculations; it was about the communication of risk and uncertainty. In the period leading up to the earthquake, the area had experienced a number of small quakes, known as seismic swarms. While experts disagree about the extent to which such swarms influence the probability of a big earthquake, there is consensus that it cannot be seen as a deterministic precursor. However, one of the experts who was put on trial – De Bernardinis, a government official, and the only one whose sentence was not overturned – had said in an interview that such a swarm was actually a good sign. He claimed that swarms release energy, a statement that is widely considered scientifically false. He also called the situation 'normal', and encouraged the inhabitants of L'Aquila to go home and spend the night indoors. In a subsequent meeting of the expert committee and a related press conference, a different message was expressed, emphasising that earthquakes cannot be predicted and that in a high-risk area such as L'Aquila nothing could be ruled out. However,

De Bernardinis' statements were not refuted, nor were precautionary measures discussed. The main argument of the prosecution was not that the experts did not predict the earthquake, but that they were negligent in their assessment and communication of risk and uncertainty. Also, the main argument for overturning the six other experts' verdicts was that they could not be blamed for the statements made by De Bernardinis. In other words, the appeals court did uphold the view that De Bernardinis' statements were a decisive factor in causing casualties by influencing the behaviour of the residents so that they remained indoors (Cartlidge, 2015).

Second, the advice of the expert committee was not the only source of information available in L'Aquila. Local inhabitants had adopted a number of precautionary measures, including spending nights outdoors after seismic tremors. One inhabitant admitted that De Bernardinis' reassurances had led him to break with this tradition (Hall, 2011). A local resident, Guiliani, had been making his own earthquake predictions using radon gas levels, and he had warned that a major event was on the way. Although his ideas were controversial, they were picked up by the local community. According to some commentators, this created discussion in the expert committee and subsequent moves to silence Guiliani and avoid panic (Hall, 2011). As a consequence, public communication focused on whether or not there would be an earthquake, which drew attention away from communication about preparedness and about what preventive and precautionary measures could be taken.

There are several lessons in this case for risk communication, which requires a clear division of labour. It needs to be clear who is communicating about earthquake expectations and who makes policy decisions based on them. Experts are expected to state their assessment of the situation and avoid spinning the facts for political convenience. However, in a local context, local customs and knowledge about handling risks should not too readily be discarded as 'unscientific'. In fact, it may be counterproductive to get bogged down in discussions about whether dangers are sufficiently proven or not, at the expense of practical measures for coping with environmental hazards. Throughout this book, we will provide further examples of this difficult balancing act between sticking to independent experts' standards and keeping an open mind towards other forms of knowledge and other approaches to handling danger.

The L'Aquila case illustrates how hard it can be for experts to advise on public decision making, especially under conditions of high stakes and uncertainty. But even when nature does express itself loudly and without scientific intermediary, with environmental disasters, dwindling rain forests, or large pollution accidents, it may not automatically be clear where to locate the original cause, how to attribute responsibility, or how to identify optimal solutions, even with the help of scientists.

Environmental problems hence require more than just scientifically validated facts. They require deliberation over which values are at stake and which causal connections involve obligations to act. For example, should we blame ill-informed consumers for the pressure on the world's fish stocks, or hesitant regulatory agencies, under pressure from fisheries ministers worried about jobs in the fishing industry? Or should we blame Western consumption patterns for exhausting marine resources in southern oceans? Should the protection of fisheries involve edible species primarily, or find some balance with the protection of marine biodiversity, even if it cannot be harvested? Scientists' factual observations are mixed with questions about what should be valued in the environment, or about who has access to environmental resources, and when and how. Depending on how we answer such value-laden questions, different facts become relevant for our understanding of the environment, perhaps involving different forms of scientific expertise, or facts from sources other than professional scientists. Even if everybody paid attention to scientists all the time, experts would not be able to answer these questions just by providing facts based on the standard methods of their field.

To complicate matters further, in many environmental issues the facts are less than absolute. High levels of uncertainty are rife in environmental affairs: there are things we do not yet know, things that are too costly to measure precisely, scientific disagreement about how best to measure or model the environment, and chaotic processes that are inherently too indeterminate to model. In addition, the environmental sciences are multiple: a toxicologist will have a different approach to pollution than an ecologist or an environmental engineer, even if they do agree on facts. Also, scientists are not the only actors in society who speak for the environment: environmental movements, governments, companies, citizens, and the media also make claims about what is going on. Citizens concerned about local soil pollution may present worries over shaky facts presented by official sources. Or maybe these same citizens have dug into the technical details of pollution measurement, hired expertise of their own, and are about to make a strong case that goes against the dominant understanding of soil pollution risks. (Concerned citizens will do that, if they have the resources.)

A multitude of voices, partly contradictory and often displeased, is the norm rather than the exception in environmental issues, especially with new and emergent concerns. Shale gas drilling ('fracking'), river and sea pollution by (micro-)plastics, and the socio-environmental consequences of biofuels are examples of issues that spark fierce debates. The resulting controversies can land even the best-intentioned environmental scientist in dire straits. In the case of biofuels, many environmental scientists had expected environmental NGOs to support rapid introduction, only to find that NGOs expressed serious doubts, questioning the promise of carbon emission reduction and the predicted impact on food production from biofuels. These scientists faced unexpected levels of distrust, differing understandings of the problem (livelihood concerns, rather than carbon cycling), and suspicion over affiliations with the oil industry, as well as doubts over presented facts and challenges to established methods. Under such conditions, simply presenting more facts or insisting on the validity of standard methods rarely silences the critics. In issues that touch on peoples' concerns and livelihoods in such fundamental ways, the job of an environmental scientist is truly complex.

Our key message in this book is that the simple juxtaposition between 'sound, scientific fact' and value-laden advocacy provides inadequate guidance for environmental professionals. In addition to sound scientific work that lives up to expert standards, environmental professionals need to understand how to recognise variation in problem definitions, understand sources of distrust in policy or expert institutions, and appreciate the importance of acknowledging and communicating uncertainties. This requires an understanding of the variety of roles experts play in environmental issues, beyond merely providing accurate facts, however crucial these may be.

1.3 The Book

This book will help you prepare for, or reflect on, the challenges of a job in the incendiary world of environmental affairs. Specifically, it will suggest ways to prevent surprises, identify and recognise deeper sources of disagreement over environmental issues, communicate about uncertainties and limitations of knowledge, and position yourself in times of environmental controversy. This book is not about methodology or about how to get your facts straight. Rather, it is a book that explains how facts and values are entwined and how controversy can often arise not just from disagreement about the facts, but also from debate about which facts are to be considered relevant. Our main goal is to offer guidance on how experts can play a productive role in the governance of current environmental challenges, while respecting the diversity of perspectives and knowledge claims and taking into account the concerns of a democratic society.

While these goals are relatively practical, we offer more than just hands-on instructions. Our book is rooted in decades of social science research on the role of science in environmental decision making, mostly connected to the scholarly fields of science and technology studies (STS) and policy studies, but also public understanding of science and the social science side of environmental studies. Our account is informed by the theoretical reflections this research has produced: rather than simple dos and don'ts, we want to convey the deeper insights of this research. Rather than rules of thumb, we want to explain the logic behind them.

We therefore take time to explain some of the theoretical debates in these fields of study, including the philosophical problem of whether science can be distinguished from other forms of knowing. We also provide pointers to relevant academic literature, thereby allowing the reader to dig deeper, or suggest materials that could be used in academic courses or seminars. However, we also put these theoretical insights to work and show their practical relevance for the work of an environmental professional. For this same reason, we combine the discussions of theory and general insight with extensive case studies to show how general patterns and processes can be observed in concrete environmental issues.

The book moves from more fundamental issues about the nature of scientific knowledge, and of how to understand problem definitions or (more generally) 'frames', to more practical suggestions for communication and organisation of expertise. It ends with a more normative discussion of expertise in a democratic society. At the end of many of the chapters you will find case descriptions that illustrate how the general patterns described in the main text may pan out in practice. We have chosen our case studies to cover a wide range of environmental matters, from fisheries to climate change or the regulation of pesticides. This is not just to illustrate the general difficulties of environmental professionals' work, but also because, as lecturers, we know the value of case stories as a 'memory hook' for general principles.

Hence we start (Chapter 2) with a more fundamental understanding of why a simple distinction between science and politics, and between science and other forms of knowing, provides insufficient handles for environmental advisory work. Our goal for this chapter is not to deny the unique value of scientific work, nor the importance of sticking to scientific standards, but to explain why referring to such standards alone fails to resolve debates, especially if environmental issues are heavily contested. The problem of how to define science is much more than finding a smart phrase to express its unique features, and it raises fundamental questions about how policy makers and citizens can trust expert knowledge, and about what knowledge will come to bear, and how, on public deliberations.

Chapter 3 makes these complications more concrete, with the introduction of *framing*. This concept helps us to comprehend and analyse differing understandings of environmental problems, which may point to varying sets of facts or assessments of uncertainty. As participants in an environmental debate define problems differently, or have radically different understandings of how the environment or environmental institutions work, complex misunderstandings may develop. A case discussion of climate change sharply illustrates these points. The chapter suggests ways for environmental professionals to recognise contrasting frames, and suggests that an awareness of frame differences can be a first step to more meaningful deliberation.

Chapter 4 focuses on environmental controversies and the particularly difficult dynamic that arises when environmental issues become polarised or if disagreement 'heats up'. These are the circumstances in which environmental professionals are confronted with hypercritical opponents who may scrutinise and question every assumption (as illustrated in the 'Climategate' case), or even accusations of partisan positions, potentially accompanied by heavy emotions (as becomes clear in the case study of shale gas or 'fracking').

Chapter 5 discusses the 'The Limits to Knowledge', providing an overview of forms of uncertainty and how these play a role in environmental deliberations. The chapter provides pointers for how uncertainties can be identified, as a tool for both clarification of environmental debate and for problem finding. Uncertainties are illustrated for the case of the flower industry, which describes the debate over what constitutes an adequate understanding of pesticide hazards.

This brings us to the next logical question of what constitutes usable knowledge, the central question of Chapter 6. If society expects more from environmental professionals than just a stream of facts, and if a stream of facts alone clearly does not help to advance environmental decision making, then what more can be expected? The chapter describes how environmental professionals can better understand the context in which their knowledge has to operate, assisting with providing timely, trustworthy, and relevant knowledge in full respect of scientific standards. The difficulties of combining such principles in the practice of policy advice are illustrated by a description of European fisheries policy.

One way environmental experts have tried to resolve contrasting accounts of environmental issues is by speaking with one voice through 'integrated environmental assessments'. Chapter 7 describes such efforts with examples from air pollution modelling and the Millennium Assessment, but also points out the downsides of integration, as some of the wealth of cognitive diversity is sacrificed for a unison account.

Lay knowledge, the topic of Chapter 8, is an important example of such cognitive diversity. Citizen science, knowledge gathered in activist opposition to policies, and knowledge of enthusiasts (such as volunteer biodiversity observers) may present valuable knowledge that is not produced or maintained by professional scientists. However, the successful use and fostering of lay knowledge requires care and tailored institutions, to address concerns over, for example, data access or the validity of such knowledge. The chapter provides illustrations of such problems for lay involvement in botanical gardens and nature conservation.

Chapter 9 translates knowledge of these processes into a description of the kinds of tasks experts perform for environmental policy, expanding beyond the narrow and instrumental job of providing 'scientifically sound facts'. The analysis of a variety of expert roles in environmental policy is illustrated with a case about the complex debate over the advantages and disadvantages of carbon storage.

Chapter 10 investigates the responsibility of environmental professionals and their expert organisations not just to scientific standards, but also to their wider role in democratic societies. The chapter takes a more normative stance and proposes some crucial virtues for the presentation of expert environmental knowledge in society.

We sincerely hope our book will prove useful for environmental professionals, in training or as reflection on their work, and look forward to continue the challenging but exciting conversation with those of us who study environmental expertise in action.

References

Brown, T. (2012). A Chilling Verdict in L'Aquila. *The Guardian*, 23 October, www.theguardian.com/science/2012/oct/23/chilling-verdict-laquila-earthquake

Cartlidge, E. (2015). Why Italian Earthquake Scientists Were Exonerated. *Science Magazine*, 10 February, www.sciencemag.org/news/2015/02/why-italian-earthquake-scientists-were-exonerated

Davies, L. (2012). Jailing of Italian Seismologists Leaves Scientific Community in Shock. *The Guardian*, 23 October, www.theguardian.com/world/2012/oct/23/jailing-italian-seismologists-scientific-community

Fox News. (2012). Italian Court Convicts 7 Scientists for Failing to Predict Earthquake. *Fox News*, 22 October, www.foxnews.com/science/2012/10/22/italian-court-con victs-7-scientists-for-failing-to-predict-earthquake.html

Hal, S. S. (2011). Scientists on Trial: At Fault? *Nature*, 477, 264–269.

Kington, T. (2012). Italian Scientist Convicted over L'Aquila Earthquake Condemns 'Medieval' Court. *The Guardian*, 23 October, www.theguardian.com/world/2012/ oct/23/italian-scientist-earthquake-condemns-court

2

What Is Science? (And Why Does This Matter?)

WILLEM HALFFMAN

2.1 Trust Me, I'm a Scientist

What is science? The question may sound academic, the kind of question only professional philosophers of science would worry about. Not so. In fact, it is a very practical question, and its answer can be of enormous consequence. Knowledge that can claim to be 'scientific' generally carries more weight and that, in turn, can affect how people make some rather important decisions. Here are some examples:

– *What knowledge is scientifically sound enough to justify expensive climate policies?*
Climate sceptics have challenged what counts as 'sufficiently sound evidence' for anthropogenic climate change (Lomborg, 2001), and whether the Intergovernmental Panel on Climate Change should be considered 'political' rather than 'scientific' – and not always for the noblest reasons (Oreskes and Conway, 2010). See also the case on climate science in this volume (Hulme, Chapter 3).

– *What is permissible scientific evidence in a court of law?*
Is the forensic identification of handwriting sufficiently 'scientific' to be allowed as expert evidence in court? Or is it more akin to *graphology*, the dubious pseudo-science that infers emotions and mental states from handwriting? Legal challenges have effectively required judges to demarcate 'science' from 'merely experiential knowledge' of handwriting. Historically, the verdict on handwriting expertise has been by no means straightforward (Mnookin, 2001; Solomon and Hackett, 1996).

– Should creationism be taught in public schools?
Traditional religious objections against evolutionary biology relied on theology: Darwinism contradicted religious texts, which were presented as the higher authority. Modern creationists try to mobilise the authority of science by claiming that theories of intelligent design have scientific credence, or, inversely, by challenging scientific evidence in evolutionary biology. Their attempts to replace Darwinism with creationism in school curricula have challenged what counts as 'science' (Gieryn, Bevins, and Zehr, 1985), or have presented creationism as 'science also'.

Establishing exactly which knowledge can rightfully claim to be 'scientific', and hence receive the extra cognitive authority that comes with it, is surprisingly difficult. As an (environmental) scientist with the best of intentions, you may want to claim that the world should believe you because your knowledge is 'scientific', but in practice things are not so simple – even if your knowledge really is quite solid. In this chapter, we will describe attempts to resolve this problem through various 'gold standards' to distinguish solid, scientific knowledge from all the rest. We will show that there are no simple solutions. 'Trust me, I'm a scientist,' or 'we used the scientific method' are not going to convince people. This is a problem that cannot be resolved with simple analytic distinctions such as 'scientific', 'specialist', or 'expert'. In the practice of environmental expertise, facts and values are intricately connected in ways that defy a simple separation into distinct roles for scientists and non-scientists. After we clear this conceptual dead end in this chapter, the rest of the book will suggest more fruitful ways forward by describing the processes and institutions that either challenge or build the trust in knowledge.

This chapter is probably the most conceptual in the book. It tries to address some deep-rooted assumptions about science and knowledge, in order to make way for alternative conceptions and experiences. The chapter builds on many concrete examples, but in case you find it still too abstract, we recommend you proceed with the rest of the book and return to this one later. Although this conceptual problem logically precedes the rest of the book, reversing the order may be more instructive for some.

2.2 The Reputation of Science and Its Uses

Even with good arguments, it can be very difficult to come to an agreement on what exactly may count as scientific. If we momentarily suspend our urge to decide who is right and who is wrong, then we can at least establish that whether something is 'proper science' can be of grave importance, potentially raising a lot of agitation. Is this vaccine really safe, and is contrary anecdotal evidence 'not scientific'? Should we take the knowledge of local people into consideration for regional planning decisions, or stick with science-based knowledge only? Can patients participate in an advisory body on disease management, or only scientific experts? Should we stop public funding for a field of research because it is not universally considered 'properly scientific'? If you want knowledge to be part of school curricula, or of (legal) decisions, political deliberation, or public research funding, it seems to help a lot if you can claim that this knowledge is 'scientific'.

In most societies, science tends to have a considerable reputation. In spite of populism, science is still more trusted to establish factual truth than other institutions: if we want to know what is safe or healthy, what the state of our environment is, or what our options are for a sustainable future, we generally trust scientists over companies, governments, or social movements. Science has cognitive authority: if science says it is true, then it must be so – or at least highly likely, to most people. The label 'scientific' acts as a certification for the reliability of knowledge.

Even though science's cognitive authority holds in general, there are important qualifications. First, not everybody trusts science to the same degree. Scientists and more highly educated people tend to trust science more than the general public, the level of trust varies considerably between countries, and there seems to have been some erosion of trust over time (De Jonge, 2016; Gauchat, 2011). Second, there are good reasons not to trust science blindly. A deeper understanding of science may also lead to criticism of some controversial techno-scientific endeavours, such as genetic modification or animal experimentation. The history of eugenics, abuse in human experiments, and stories of scientific fraud remind us that our trust in science should not be too unconditional. Science was never perfect.

The cognitive authority of science is symbolised in heroic stories that we tell our children and students. We tell tales of how Galileo Galilei was almost burnt at the stake for a better model of the solar system, or of the immeasurable genius of Newton, and Einstein. We even keep Einstein's brain in a jar, because some people thought it might reveal the secret of his brilliance (Abraham, 2004). If you grew up in France, then you will have learnt to celebrate Louis Pasteur for conquering superstition and bringing us vaccines (Geison, 1995). In environmental science, you probably know of the scorn and ridicule that befell Darwin for his theory of evolution (Desmond and Moore, 2009), and of Rachel Carson's courageous struggle to show how organochlorine pesticides threatened our birds (Carson, 1962).[1] We honoured these people with statues in parks and market squares, wrote their biographies, and put their faces on postage stamps or bank notes. Modern culture celebrates science (even if it does not always fund it accordingly).

Let us be clear about this: the cognitive authority of science is well earned. Science has an impressive record of achievement. Just think about our increased life expectancy (at least in the richer parts of the world), the increased productivity of agriculture, our environmental awareness, the knowledge of our distant past, and of our place in the cosmos. The authors of this book would not turn to quackery and superstition if they got ill, but to tested medicine. We want proper, certified science to assess the quality of our drinking water, and we trust our life to the wonders of engineering and aerodynamics whenever we step into an airplane. Even though we may point out many complications, the cognitive authority of science is not achieved through empty rituals, but is based on the accumulated improvements to the human condition of the past, using carefully honed methods and institutions.

Even though we may be critical at times, daily life in a technological society would become impossible without some level of delegation – that is, trust in scientists' assessments of matters we cannot figure out for ourselves. For the safety of our drinking water, food, trains, or pills, we rely on regulatory standards that are at least partly based on scientific knowledge. If we had the talent and the time to study for ourselves the toxicology, food science, railroad engineering, or pharmacology involved, we might theoretically be able to verify the solidity of the knowledge involved. In practice, citizens, policy makers, companies, and judges rely on delegation: we hesitatingly trust the specialists who assess such safety issues, and expect that the science involved will be impartial, sound, and considerate. Thus, identifying some knowledge as 'scientific' is one way to manage the division of labour in a complex society and delegate the detailed assessment of knowledge claims to science-based experts.

Non-scientists also try to rely on science's cognitive authority to make claims about what is going on in the world, or about what should be done. They try to strengthen their statements by claiming that they are 'scientific'. They may claim it is 'scientifically proven' that certain foods are healthier than

---

[1] Often, the heroes in the stories are physicists and male, reflecting older historic biases. The stories also over-expose the positive aspects of such heroes. After all, Galileo recanted at the last moment, part-time alchemist Newton obscured the contributions of his great competitor Robert Hooke, Pasteur won his battles at least as much through clever theatrics as through brilliant experimentation, and Darwin's fear of religious backlash made him delay publication for more than a decade.

others, or that there is 'scientific proof' for climate change. Or you may claim your company's product was 'scientifically proven' to be superior to a competitor's. As a policy maker, a consultant, or a company you might be able to use science's reputation in support of your activity or decision. It is not just scientists who claim superior credibility, but also those who use scientists' knowledge.

The cognitive authority of science can also rub off on your activities. To do that, you would have to be 'just like science': you would need grounds on which to claim that what you are doing is part of, or similar to, the scientific endeavour. For example, you could argue that, just like science, your profession is taught at a university, with academic reflection, resulting in an academic degree that is recognised beyond your local private college. You might want to point out that what you are doing is based on recognised methods, certified by peers, and widely accepted; that you use a laboratory, mathematics, falsifiable theories, or some of the other things that people recognise as characteristically scientific. If you can claim that a practice, a measuring technique, a device, a statement, a person, or an organisation is 'scientific', then you implicitly mobilise the impressive legacy of science in support of your activities.

However, there is also a catch. Many people have tried to appeal to the authority of science in the name of various causes, but on shaky grounds. For example, throughout the nineteenth century, phrenologists claimed they could deduce mental capacities from the shape of human skulls or bodies, using dubious biometry to justify inequality, sexism, racism, and even slavery (Desmond and Moore, 2009; Gould, 1981). In recent times, governments and companies have appealed to scientific arguments to downplay the hazards of chemical pollution, nuclear installations, medicines, or food contamination, such as in the case of 'mad cow disease', often to avoid costly safety measures or avoid inconvenient upheaval. Similarly, we would not want to allow any quack doctor, who's out for a quick profit, to claim scientific credentials. It is not because you appeal to the cognitive authority of science that you auto-matically deserve to shelter under its umbrella.

There is also the opposite problem: if you reserve cognitive authority too strictly to only the 'hardest' of sciences, then you may discard a lot of valuable knowledge (see Chapter 8, this volume). Indigenous peoples of the Amazon may not read the *Journal of Pharmacology*, but they can point pharmacologists to promising substances through their intimate knowledge of forest plants (Minnis, 2000). Patients with chronic diseases, such as diabetes or HIV/AIDS, may use their knowledge of daily disease realities to inform better care plans, even if their knowledge is not based on double-blind clinical trials, the somewhat overrated 'gold standard' of medical research (Epstein, 1996).

Field biology enthusiasts, such as experienced birders or lifetime bryologists, may gather extensive biodiversity knowledge that we could never afford through professional ecologists (Ellis and Waterton, 2004; Lawrence and Turnhout, 2010). Including too much under the umbrella of science's cognitive authority may present the risk of admitting nonsense, but, clearly, allocating cognitive authority too strictly risks discarding a lot of valuable knowledge.

It would be convenient if there were a clear criterion, an infallible yardstick that would allow us to separate true science from quackery, reliable knowledge from dubious superstition and pseudo-science, reserving the accolade of science's cognitive authority only for what truly deserves it. Scientists seem to recognise good science when they see it, so surely there must be some logic to their assessment? Can't we devise a practical definition of science, or at least of reliable knowledge, and then use that to decide who and what is 'in' or 'out'?

2.3 Science as a Particular Way of Reasoning

It is surprisingly hard to come up with a clear and universal definition that covers everything that you would commonly call 'science'. It is easy enough to list recognisable characteristics: science tends to perform empirical studies and accumulate facts, gathered through experiments or at least using well-established methods of systematic observation or registration, integrated through general theoretical notions and understanding, which are then tested against further facts, in a quest for general laws and regularities, all in a spirit of healthy scepticism towards tradition and accepted belief. With such a list of characteristics, science appears as a particular way of gathering and accumulating knowledge that seems more systematic and objective than ordinary forms of knowledge. We could argue about the details, but this is roughly the descriptive definition of science that scientists are taught during their training.

However, if we try out our list of characteristics on *specific forms* of scientific knowing, then it quickly becomes hard to make all of them fit. For example, we may think of experiments as the epitome of scientific research, but some sciences are not very experimental at all, such as astronomy or sociology. Large parts of climate science work with computer models rather than experiments or systematic empirical observation, as does much of economics. We could follow the Anglo-Saxon usage of the word 'science' and restrict its meaning to the *natural* sciences only, as opposed to the 'social sciences' (in contrast to the tradition of Continental Europe, where 'science' generally includes all academic knowledge). Even if we were to reserve the accolade of 'scientific' to the natural sciences only (which seems to discard rather a lot of valuable knowledge), this still does not mean that all 'science' is experimental. In fact, some highly respected sciences, such as mathematics, are not very empirical at all. While some sciences are occupied with testing generalisable laws, fields such as taxonomy deal with nomenclature and classification instead.

Even a concept such as 'objectivity' is more problematic than you might expect. It might mean 'without value judgement', or 'without any human intervention', or 'checked and agreed by many people', or 'accurately representing the world' – all of which become quite complicated in any concrete example of research (Daston and Galison, 2007). In some sciences, 'objectivity' is treated with a lot of circumspection. For example, anthropologists recognise that their presence in the communities they study, or their own inescapable cultural biases, affect their analyses. To say that mathematics, anthropology, taxonomy, and astronomy are 'not really sciences' would not only be counterintuitive, it would also deny these obviously valuable forms of knowledge the cultural legitimacy and support provided by the label 'science' (or 'social science', if you insist).

Inversely, some forms of knowledge are empirical and theoretically integrated, but would not be seen as science by most people, such as astrology. Astrology uses astronomical tables and relies on the mathematics involved, consists of theories about how planets influence people, and involves established methods for predicting the future, such as a birth horoscope. It learns and improves (at least by its own standards) – for example, it incorporated the planets beyond Jupiter as they were discovered. Although we do not want to suggest that astrology is a science, it is important to realise that even philosophers of science disagree about exactly *why* it is not a proper science.

In addition, some forms of knowledge operate somewhere in the grey zone, such as herbal medicine or acupuncture, with debated evidence and controversy about what counts as reliable evidence. Advocates and practitioners of 'alternative' sciences may engage in fierce and bitter debates with mainstream scientists that can last for decades. Some of these debates result in eventual rejection, such as the case of phrenology: the study of how skull shapes express personality eventually became 'unscientific'. In other cases, the 'unscientific' outsiders eventually are accepted into the fold, as in the case of plate tectonics, a field of study that took decades to achieve recognition. Some disputes never quite seem to get fully resolved, as with homeopathy, which has had a precarious position on the fringes of science for decades.

Philosophers of science call this the problem of demarcation: what criterion could we use to demarcate true, reliable scientific knowledge from other forms of knowing? A famous attempt, still quoted often by scientists today, was made by the philosopher of science Karl Popper: science works with falsifiable hypotheses. Rather than looking for facts that can confirm theories about the world ('verification'), the true scientist should articulate claims about the world in such a way that they are open to challenges and then try to refute these claims. For example, we should not assume that all swans are white and then find in every observed white swan a confirmation of our belief, but rather look for black swans and try to refute our tentative theory that all swans are white. Until the day we actually find black swans,[2] we can cautiously assume that all swans are white, but we must accept that this is only a preliminary certainty – that is, the best we can do at that point in time. Popper used this criterion to distinguish science from pseudo-sciences such as astrology or occultism, which always seemed to find a way to reinterpret observations to confirm their beliefs.

More than just a philosophy of science, Popper also wanted to challenge over-commitment to ideological beliefs. With his principle of refutation, he distanced himself from Marxism, which in his opinion failed to accept observations that ran contrary to its expectations. After Marx had pointed out that capitalism undermines its own existence and therefore must lead to revolution, his followers were forced to find ad hoc explanations for why capitalism continued to survive its 'internal contradictions'. Tellingly, Popper's theory of science was first published in German in 1934, under the looming threat of Nazism, and republished in English during the Cold War in 1959 (Popper, 1934/1959). It became part of his defence of an 'open' and liberal society, against totalitarianism (Popper, 1942/1966).

While immensely influential and still quoted often, philosophers of science have raised quite a few objections to Popper's demarcation through the refutation principle, especially when confronted with the actual practice of research. For example, when a refuting observation occurs, it is not immediately clear *what* should be rejected: is the theory wrong, or did the experiment fail? Was the observation an outlier, or was the observer incompetent? Because we have no absolute way to determine the cause of refutations, of whether they result from error of measurement, the experiment, or the observer, it is impossible to answer these questions with certainty. This means that the application of Popper's principle will run into inevitable difficulties in practice.

How to handle refutation is further compounded by the problem of interpretation. What do we do if we find a black bird that looks like a swan? Do we say that it is a swan and refute the theory that all swans are white, or do we revise the definition of a swan to exclude black ones? (While this latter option

---

[2] Black swans (*Cygnus atratus*) actually exist, originally from Australia, but also as escaped park populations.

may seem silly, black and white swans are in fact considered two different species.) Or do we just accept that the matter remains unresolved? Thomas Kuhn famously observed that scientists regularly allow anomalies to accumulate, rather than refute well-supported theories that have proven useful. It is only when sufficient evidence accumulates that completely new approaches ('paradigms') will be considered, typically by a new generation of scientists who are not so committed to the old ways, launching a 'scientific revolution' (Kuhn, 1962/1970). An example of such a scientific revolution is the change from Newtonian to relativist mechanics, or the rise of evolutionary theories in biology. Popper's aim of falsifying hypotheses is hence a principle that only seems to work in the context of stable theoretical assumptions, defying the idea that refutation can effectively challenge fundamental worldviews, as it was intended.

Philosophers have observed that, in practice, scientists use a rich mix of epistemological principles in their research, sometimes aiming to refute hypotheses, sometimes generating new theories through observed confirmation, or even by purely deductive reasoning. If we ask them which philosophy of science they use, they answer with an eclectic mix of principles of opposing schools in the philosophy of science (Leydesdorff, 1980). This does not mean that 'anything goes': just because there is not one universal method in science, it does not mean that there are no valid methods at all (Feyerabend, 1975). But it does mean that we have to understand and appreciate that research fields have differing, evolving standards of what constitutes valid and solid research.

Hence, it is still possible to criticise some knowledge for being unfounded, or to question research in terms of its own or even neighbouring standards (Fagan, 2010). For example, paleo-climatologists may have meaningful objections to global circulation climate modelling on the basis of completely different principles for how to generate reliable knowledge: from empirical observation of traces left by long-term climate change, versus computer models based on physical laws and current measurements (Shackley and Wynne, 1996). Or we may criticise toxicologists if they do not stick to their own risk assessment methods to judge the hazards of chemical substances. Once standards have been established for how to assess pesticide hazards, it becomes possible to check whether actual assessments live up to them (see Van Zwanenberg and Millstone, 2000). There is no need to give up knowledge standards completely, but such standards tend to be much more specific than a universal 'scientific method'.

Popper was not the only one to search for the ultimate demarcation criterion. The demarcation problem was a key concern for an entire generation of philosophers of science, but none ever came up with a solution that remained universally convincing (Callender, 2014, 46). The idea of a universal 'gold standard', '*the* scientific method', or a definitive criterion to distinguish all scientific from non-scientific reasoning, is a theoretical project that never seems to work in practice. To determine which chemicals are dangerous to the environment, how to assess the effectiveness of pharmaceuticals, or what contributes most to climate change, we still need to go through lengthy deliberations over what counts as sufficiently reliable knowledge. While refutation is one useful principle to do so, it does not definitively resolve the problems of citizens, policy makers, and judges who have to distinguish valid from questionable knowledge in order to decide on appropriate courses of action, especially under conditions of polarised conflict or expert disagreement.

2.4 Science as a Particular Way to Organise Knowledge Creation

We could try to define science not as a particular style of reasoning or generating knowledge, but as a particular way to *organise* the generation of knowledge. Maybe there is something unique about the *social* structure of science, rather than its *cognitive* structure. This shifts the attention away from how scientists think to how they work, cooperate, and deliberate together.

Thus, science could be seen as a scholarly activity, relating new research to a previously accumulated body of knowledge, but in a unique format. This knowledge is openly shared among a community of researchers, open to common scrutiny and debate, and organised via peer review and a particular system of scientific publishing and communication. We then understand science as a unique set of organisations (laboratories, journals, conferences), social practices (peer review, open discussion), and shared values. For example, researchers like to present science as disinterested, neutral, unbiased in political or cultural quarrels, based on facts, objective, and universally true, irrespective of country, race, gender, belief, political convictions, or other particularistic categories. Even if the methods of science are diverse and complex, presumably these shared attitudes and organisations can distinguish it from other activities in society?

2.4.1 Unique Norms and Values?

Sociologist Robert Merton argued that science is set apart because of its unique combination of four core values. According to Merton, the norms of science dictate that knowledge should be openly available and *Communally* shared, and should not be tainted by particular characteristics of the scientist (gender, nationality, age, race, etc.), but be *Universal*. The production of knowledge should be done in a spirit of *Disinterestedness*, where private advantages should not interfere with the quest for new knowledge. Last, knowledge should subject to systematic and critical scrutiny by peers through *Organised Scepticism*: not just random criticism of every detail and assumption, but a reasonable and collective questioning of scientific findings, as can be found in peer review. He expressed this 'ethos of science' with the acronym CUDOS, with a pun on *kudos* and the importance of recognition in the reward structure of science. Only science would have this unique combination of norms, providing a social rather than cognitive demarcation criterion. Merton originally formulated the normative structure of science in 1942, with concerns similar to Popper's about the totalitarian repression of science under Nazi and Stalinist regimes (Merton, 1973 [1942]).

Many scientists would agree that the CUDOS values stand for important norms, and invoke them to call each other to order. A recent reappraisal of CUDOS has shown these values to still be relevant, especially in a critique of a science that is increasingly operated as a business (Radder, 2010). For even if these norms may be invoked and recognised by scientists as valuable ideals, the reality of research is often very different: knowledge is held back until publication priorities can be assured; rivalries between laboratories, personalities, and even countries lead to particularistic research choices and unfair opportunities; research investments are steered by commercial considerations or military advantages that require secretive knowledge rather than communal sharing; and careers are built on commitments to theoretical premises or approaches rather than sceptical self-doubt. (Such research commitments and identities are woven into how scientists are trained and rewarded, and are no less particularistic than commitments to economic or political interests.) Furthermore, governments increasingly expect scientists to affiliate with companies or government agencies to guarantee practical benefits from research. If we remove all interests from science until we have truly 'disinterested' research, there would not be a lot of research left. Our list of values may express an inspiring and acutely relevant ideal, but it is not a very useful yardstick to distinguish true science from more questionable knowledge creation.

This has led to the objection that daily research practice is often guided by more dishonourable 'counter-norms': science that is secretive instead of communal, particularistic instead of universal, interested instead of disinterested, and dogmatic instead of open to scrutiny and organised scepticism (Mitroff,

1974; Ziman, 1994).[3] There is even some evidence that the increased commodification of science is becoming accepted by scientists as the new state of affairs, altering the ideals of science (Macfarlane and Cheng, 2008). In any case, whether or not the traditional ideals of science still carry weight, they do not provide a solution to the problem of what we should trust as 'true science', even if they might provide some guidance as to the kind of research we could question. Evidently, supporting beautiful ideals does not automatically produce reliable scientists.

2.4.2 Peer Review

What if we take a procedural perspective to the demarcation problem? What if 'science' is simply what scientific organisations do? For example, 'scientific' is all that is published in peer-reviewed journals, in which anonymous reviewers review papers prior to publication. The quality checks of peer review increasingly include grant proposals, and even career advances. In fact, just as with Popper's refutations, scientists and their organisations sometimes refer to peer review as the ultimate criterion to distinguish the reliable from the unreliable. To underline its importance, they sometimes refer to its origins in seventeenth-century academies of science that fostered the illustrious Scientific Revolution. There is a firm belief that peer review will ultimately weed out faulty and unreliable science: by submitting research to the anonymous, 'blind' evaluation of knowledgeable colleagues, scientists will be able to assess the quality of research without being confused by irrelevant social properties of the knowledge source, such as reputations or nationalities.

However, peer review is a varied practice, and not even unique to research. Some forms are used outside of science in the assessment of professional work, among nurses or teachers. The double-blind review system of the humanities (where names of authors are removed from submitted papers) is uncommon in some natural sciences, particularly in small, highly specialised communities where anonymity is an illusion anyway. In fact, in some areas of physics quick online access seems to be more important than pre-publication peer review. In any case, peer review originated not in quality assurance, but in censorship: review was to prevent the publication of seditious or subversive material that could undermine the authority of the king or the established social order

---

[3] Expressed by John Ziman as the PLACE counter-norms (Ziman, 1994): 'post-academic' science is Proprietary (concerned with owning and earning from knowledge), Local (e.g. in search of regional advantages), Authoritarian (as in large hierarchical research programmes), Commissioned (by states or companies), and Expert (advising with certainties rather than cautious doubt).

(Biagioli, 2002). As a system of quality guarantee, it only became common in the natural sciences after World War II, and in the rest of the sciences even more recently. This would mean we would have to exclude Newton, Darwin, and even Einstein from science, as their work was published long before peer review became a standard practice.

As a quality guarantee, peer review is under a lot of criticism. Peer review has a conservative bias, making it hard to publish ideas that go against dominant beliefs. In fact, several Nobel prizes have been awarded to work that was originally turned down by the peer review system (McDonald, 2016). Novel or interdisciplinary work can suffer from the fact that relevant and competent peers may be hard to identify. There are also indications that peer review favours established scientists, can be biased and even abused, and is subject to negligence and self-interest (Biagioli, 2002). Experiments where faulty papers were submitted for peer review revealed less than desirable levels of scrutiny, especially among the growing number of fringe, open-access journals (Bohannon, 2013). In fact, scientists themselves often appear unimpressed by a publication's approval through peer review in general, and rely instead on characteristics such as the likelihood of claims in light of prior knowledge, or the reputation of a journal, the author, or the author's institute (Collins, 2014). Without questioning the immense value of some peer-review practices, it seems highly problematic to find in peer review a simple gold standard establishing what we should and should not take to be solid and reliable scientific knowledge.

2.4.3 Laboratories and Centres of Calculation

French anthropologist and philosopher Bruno Latour and his colleagues have highlighted the crucial position of laboratories in science (or similar places where knowledge is created, such as field stations and botanical gardens), not as a way to demarcate science, but to describe the particular processes by which research operates. The laboratory is the place where scientists collect samples and measurements (even from places as far away as the tails of comets), and subject these bits of collected nature to trials and tests to register their behaviour. The registrations collected by scientists produce an enormous universe of carefully noted 'inscriptions' and calculations, ordered and re-ordered into data repositories and publication libraries. This perspective focuses on the remarkable processes of accumulation in science, showing the dazzling networks that connect scientists to the world, rather than to separate them from it (Latour, 1987).

The great feat of scientists is to re-enact the tamed nature of the laboratory, of the controlled experiment or model, out in the world again: to make 'tamed' nature repeat in the wild what it did in the laboratory. This turns the laboratory into a powerful instrument to change the world in which we live – as, for example, when scientists use tamed viruses to produce vaccines (Latour, 1983, 1993; Latour and Woolgar, 1979). If we use the word laboratory in a wider sense (as is more common in France, where you can have a 'laboratory' of sociology without test tubes or microscopes), we can see how such 'centres of calculation' operate throughout the history of science and its diversity. Researchers collect survey results, exotic plants in botanical gardens, or naval explorers' sketches of coastlines for accumulation in colonial atlases (Callon, 1986; Jardine, Secord, and Spary, 1996; Law, 1987). Research becomes a particular way to gather, control, and accumulate knowledge.

This view of science stresses the practical work of scientists, rather than lofty theory or interpretation, a perspective that scientists sometimes see as alien or even offensive. For many scientists, it misses the deeper meaning of their work, and the systematic reasoning and argumentation it involves. Latour himself has insisted that social scientists and philosophers should not try to demarcate science. In trying to describe science (rather than perform it), we should never draw boundaries, because we run the risk of overstating the rigidity of distinctions that would hide the close connections between science and society. Hence, we should strive never to make boundaries harder than the actors would (Latour, 1987). In more recent work, he has stressed that science uses particular modes of operation, different from law or religion, but insists that theses modes are complex and cannot be reduced to simple demarcation criteria (Latour, 2013a, 2013b).

2.5 The Diversity of the Sciences

2.5.1 The Family of Sciences

Attempts to distinguish science from other forms of knowledge aim not only to find a hard boundary, but also to gather all of science under one umbrella: they insist on the unity of the sciences. Such a project may make sense in historic circumstances that threaten free inquiry, such as religious intolerance, totalitarian regimes, or waves of mass superstition, but there are important disadvantages too. A universal criterion for what is and is not science may threaten the wealth of cognitive diversity in the sciences. It entails the danger that one way of doing research becomes the gold standard, imposing awkward or even inappropriate criteria on other sciences. Throughout the twentieth century, experimental physics would count as an epistemological ideal, the ultimate 'hard science', and many other fields of research abandoned older styles to shape themselves in its image. These days biomedical research has become a powerful template, even though their approaches may sit awkwardly with field research, anthropology, or information sciences. In addition, a strict demarcation runs the risk of discarding too much knowledge, knowledge that is practical or true, but not experimental and peer reviewed, such as patients' knowledge of how to live with the daily realities of a chronic disease. We will return to the latter problem in Chapter 8, on science and lay knowledge.

If we look at the daily operation of science more closely (scientists at work in their laboratories, climate modellers behind their computers, social scientists interviewing people), we can observe how diverse the practices of these researchers truly are. Perhaps it is not so strange that we have such a hard time finding simple demarcations for knowledge practices that are actually so varied: calculating, modelling, theorising, counting, experimenting, interpreting; in diverse places such as observatories, laboratories, museums, botanical gardens, observation stations, space stations; producing various outputs including reports, research papers, technologies, public statements, and lectures. Some researchers use extremely expensive equipment, others rely on libraries, databases, or just their desktop computer.

There is something particular about the idea that there is a shared criterion in all of these ways of producing knowledge. One way to describe such a criterion metaphorically is as a search for the largest common denominator: the most meaningful characteristic we can find which all the sciences all have in common. This approach to the problem of demarcation is an analytic one: we try to identify essential characteristics in the diversity of the sciences. In itself, there is nothing wrong with subjecting the sciences to such an analysis. However, a problem arises when we inversely start to use this essence as a criterion to judge who and what really belongs, and who and what does not. The project then acquires a nasty, fundamentalist character: only those who have the pure trait can claim to be true scientists, while the others will have to be inferior mongrels, expelled from universities, or tolerated but with a lesser status. Turning back an essential, analytic definition of the sciences as a demarcation criterion then becomes a brutal essentialism used to expel some forms of knowledge.

Ultimately, attempts to find some pure essence that all the sciences have in common (hence the term 'essentialism') have failed to provide a definitive and universal answer, even though some are still looking. Perhaps it is more accurate to think of the sciences as an extended and somewhat rambunctious family. This family shares some resemblances and a number of more or less common characteristics,[4] such as empirical work, theoretical thinking, systematic testing, and peer review. However, there are some eccentric uncles who lack some such features. There are also some distant nephews and cousins who married into the family, but are not yet fully accepted by some of the grandmothers. There is also a lot of fuss about who actually rules the roost and who represents the purest family traits. Ultimately, this family is a rowdy, fuzzy bunch, with ever-contested membership. The search for demarcation is a search for a clear criterion for who belongs in the family, who is allowed to share the inheritance. Favouring some characteristics over others runs the risk of losing some valued members, while a happy-go-lucky open-door policy means some shady suitors could take off with the family heirlooms. The family of the sciences seems too complex, extended, and dynamic to resolve its membership and proper customs once and for all.

However, to simply state that 'it's complex' is too easy and defeatist. Even though it is complex, there are ways to identify different branches in the family. Rather than one homogeneous 'science', we can try to identify more specific family branches to see how they establish the validity of knowledge claims.

2.5.2 Styles in Science

In an attempt to celebrate, rather than condemn, science's diversity, some historians of science have suggested a series of different scientific *styles*. A style is an encompassing way to express different approaches to research, involving much more than just different methodologies, but also what scientists see as ideals for knowledge (a model, a system, a law, a solution, a cure), or what they consider valid ways to establish solid, reliable knowledge (deduction, modelling, experimentation, field trials). Specific disciplines tend to have a preference for certain styles. In medicine, random control trials are often considered the gold standard for research, but other fields put more trust in computer models, or insist on laboratory experiments, theoretical deduction, or systematic observation. In some cases this is the result of practical impossibilities, such as experiments in astronomy. In policy sciences, semi-controlled field experiments once held the promise of a hard standard to test the effects of policies or other interventions, but they failed to deliver the solid conclusions they promised and are now used sparingly. Economists, on the other hand, put a high credence in models. We can also see different styles at work in how

---

[4] The family resemblance is a metaphor used famously by Wittgenstein: even though not all members of a family may have the same nose, there may be a set of features most of them share, so that family members are still recognisably similar (Wittgenstein, 1953).

knowledge is organised: field biologists who look for new species and try to systematise them in the nomenclature of taxonomy do something quite different from ecologists who try to model how predator–prey relations affect the size of populations over time, or from molecular biologists trying to identify the processes of gene expression, even though their findings may be relevant for each other.

The precise list and number of styles distinguished varies between historians of science (Kwa, 2011; Pickstone, 2000), but our point here is more to show how styles can help to identify diversity. For example, the taxonomical style is rooted in biologists that travelled along with colonial expeditions, collecting plants and animals from exotic places and bringing them back to botanical gardens, where they could be systematised and grown, and then redistributed for horticulture or commercial agriculture. The experimental style, in contrast, does not rely on collecting in botanical gardens or aiming to systematise species, but on laboratories, where the experiment is the crucial way to know nature. Its ideal is not so much systematisation, but to find general laws and regularities. In a technological style, the objective is to develop working apparatuses or procedures, whereby 'working' is more important than precise and exact understanding of underlying laws, which could come later.

Over time, styles have fallen in and out of favour. For example, at least up until the mid-nineteenth century, research in biology predominantly followed the taxonomical style: Linnaeus, Buffon, and even Darwin were fierce collectors and systematisers. Now, this style of biological research is no longer considered very 'hot' and is overshadowed by experimental research, especially in molecular biology. Many of the painstakingly gathered specimen collections have been discontinued and in most academic biology curricula, taxonomical knowledge is no longer a priority. There is a risk in such trends, of course, as old knowledge may suddenly become relevant again, for example as biologists learn to use the great collections to study climate change or ecological restoration (Bowker, 2000).

2.5.3 Disciplines

Another way to characterise the science family is as a collection of disciplines: biology, physics, geology, sociology, economics, etc. Disciplines are divisions and departments of science that structure research, with many originating in the nineteenth century. Disciplines organised the knowledge creation process in specialised publications (disciplinary journals), scientific societies and their conferences, handbooks, and degrees. The punitive connotation of the word 'discipline' is not coincidental: a discipline defines the masters students should emulate, the founders of the field, prescribes what methods and theories should be known and can be used, and how to apply them correctly. During education in a discipline, students go through strong socialisation processes. Respected and often charismatic professors and their handbooks infuse their students with a strong ethic of what is and is not proper research, of what is 'the scientific method'.

Disciplines may even prescribe correct language and formulations, sometimes in radically divergent ways. For example, whereas physics will require the researcher to be completely absent from the narrative of the scientific publication by using passive tense ('an experiment was conducted' rather than 'we conducted an experiment'), in anthropology the invisibility of the researchers in the narrative is generally considered a serious faux pas. Whereas the physicist will respect mathematisation as a convincing way to confirm theoretical arguments, the anthropologist is taught to question the assumptions and interventions needed to count anything in the first place (Verran, 2001). Hence, the criteria for what is and is not proper science differ between disciplines.

The disciplines have lost some of their status as the main building blocks of science. Their rigid separation of knowledge into distinct departments and separate communities has led to the accusation that they have thereby prevented scientific breakthroughs. Advocates of interdisciplinarity pointed to the crucial discoveries in genetics as molecular chemists entered biology, or the rise of cognitive sciences from the cooperation between disciplines such as psychology, linguistics, and biological neuroscience. Another important argument for interdisciplinarity was that disciplines tended to turn inward, to focus on research problems defined by the discipline, at the expense of questions posed by problems in society. To really contribute to problems presented by urbanisation, pollution, climate change, or health, cooperation of different disciplines would be necessary (also see Chapter 7).

Nevertheless, disciplines are still relevant categorisations for many researchers and important frameworks for organising research. For example, most universities still offer disciplinary degrees, with separate departments or faculties for 'biology' or 'chemistry', and there are still disciplinary societies, journals, and handbooks. Yet there are now more networks and communities working in various interdisciplinary fields, such as environmental studies, climate sciences, and science studies. In fact, differing disciplinary notions of what constitutes an interesting question or an acceptable way to answer it are some of the key challenges of interdisciplinary cooperation. For the same reasons, it can be hard to establish what constitutes reliable knowledge or proper ways to do research in interdisciplinary fields, which can lead to extensive debates and disagreement.

2.5.4 Fields and Specialties

Most scientists identify with a 'field' of research, with a community of researchers they consider peers, a set of research problems considered promising, a set of methods and theories (possibly in competition) to address these problems. Their relevant environment is not all of science, but their particular 'field'. It is the field that defines methodological standards, provides counter-arguments, and identifies pertinent questions. Scientists meet their field at specialised conferences, as they publish in specific journals, where they referee each other's papers. The idea that the research field defines the relevant environment for the validation of knowledge is a principle in US law, and has even been entertained by philosophers of science.

Nevertheless, it is not easy to distinctly map out fields of research. One can use citation relations between researchers or between journals: the more often they cite each other, they more they should be related. Thus, the frequency of citations can be used as a measure of distance. However, these maps are highly stylised representations of the structure of science and should not be seen as realistic 'maps' of the country of science. What is the relevant field varies strongly with the position taken in research: depending on your specific research, the field environment may look very different. In addition, this does not resolve the problem for public policy of how to establish the proper set of scientists to provide or assess knowledge, as different fields may have relevant knowledge for the issue at stake. This has been a hot topic of debate in the discussion around the attempt to demarcate who has relevant expertise.

Since scientific research tends to be performed and judged in highly specialised communities, Collins and Evans (2002, 2007) have argued that we should maintain a distinction between insiders and outsiders to such communities. Outsiders may still have relevant experiential knowledge or other expertise that can contribute to the specialised knowledge of insiders, and may to some extent even understand this knowledge, but Collins and Evans argue that only insider specialists have the genuine ability to assess the detailed intricacies of a research field. Thus, they tried to return to the project of identifying who can meaningfully assess the truthfulness of scientific knowledge, attempting to find demarcation criteria not for science in general, but for communities of pertinent experts. Their particular concern was that the growing attention given to local or experiential knowledge (people who know about their forest, patients about their disease) or 'lay expertise' could lead to the idea that everyone's knowledge counts for the same, with a resultant loss of any demarcation at all.

Their defence of insider expert knowledge has raised a storm of criticism (Jasanoff, 2003a; Rip, 2003; Wynne, 2003). One strong counter-argument is that reference to specialist knowledge does not really allow citizens, policy makers, or even other scientists to establish which specialists have the legitimate claim over a specific topic, especially if there are competing expert groups. The pre-eminence of specialist knowledge may be a strong argument in the context of Collins' favourite example, the study and detection of gravitation waves, where there is a stable topic of research defined by a long-standing group of specialised physicists (Collins, 1975, 2014). However, the suggestion that we must turn to the specialists falls short if we look for answers about the economy, nature conservation, or the environment. Not only are there competing fields of expertise on such topics, but for many such topics there is no clear definition of what the problem is. Should we protect rare species or habitats? Nature with or without people? Complex issues such as climate change can be interpreted as an economic problem, a problem of global equity, a matter of planetary engineering, a problem of adequate prediction, of attitudes towards the environment, of sheer irresolvable conflicts of interest between nation-states, and so on. With each of these interpretations or 'frames', different experts may claim to have pertinent knowledge. Collins and Evans put aside this framing as a different problem, but it is hard to see how this side of the coin can be separated from reliance on well-demarcated specialties. We will return to the problem of framing, as well as to lay expertise, in Chapters 3 and 8, respectively.

There is a more serious challenge to the idea that a community of experts can always dependably guarantee the quality of its knowledge, providing solid knowledge that outsiders can rely on. Even communities of experts may develop blind spots or fall victim to collective bias. There are examples of expert communities that have relied on assumptions that remained untested for decades – and even longer. For example, the idea that all human fingerprints are unique goes back to a conjecture by biometrician Francis Galton in the nineteenth century. Throughout the twentieth century, experts developed principles of fingerprint identification and comparison, using rules of thumb that grew out of practice and experience. Fingerprinting acquired such a solid reputation (in the law, but also in popular culture) that when DNA analysis was introduced it was often called 'DNA fingerprinting', suggesting it was equally reliable. Legal challenges of partial prints and ambivalent evidence, as well as careful reconstruction of the development of fingerprint expertise, have shown that at least some of the assumptions commonly shared by all fingerprint experts deserved closer examination (Cole, 1998). Experts and scientists are not immune to groupthink or collective bias, and the fresh but forceful look of an outsider, or of a competing field of expertise, can identify weaknesses. In spite of an understandable urge to award some reverence to specialised experts, such examples underline the importance of knowledge diversity and critical reflection by outsiders.

2.6 So How Do We Proceed?

Clearly, claiming your knowledge is true because it is scientific knowledge is problematic – not because scientific research is not valuable or a solid basis for acquiring knowledge, but because there is no unequivocal criterion to establish what constitutes 'scientific' knowledge. Even on the more detailed level of disciplines or specialties, attempts to draw a clear demarcation of what constitutes relevant expertise run into problems. Depending on how environmental problems are defined, different forms of (scientific) knowledge may become relevant. Science is diverse, and some parts of science have radically different assumptions and approaches – leading many in science studies to prefer to talk of 'the sciences' rather than a singular 'science'.

This sometimes triggers fierce responses, especially from people with a strong belief in reductionism or a shared scientific ideal, as the diversity of sciences may be misread as relativism. However, a degree of scepticism towards the cognitive superiority of science-based expertise does not automatically mean that anything is true. Diversity does not mean that there is no rigidity of scientific reasoning, or that no meaningful debate between diverse sciences is possible. It does suggest, however, that a degree of modesty is appropriate in the public presentation of science-based expertise (Jasanoff, 2003b).

In practice, scientists, policy makers, citizens, companies, NGOs, journalists, and similar actors in society have invested a lot of work in settling what we will accept as reasonable, science-informed principles to establish what counts as reliable knowledge. Over time, this results in institutionalised boundaries: anchored in rules, protocols, devices and apparatuses, communities of experimenters that share conceptions and skills for 'correct' experiments, as taught at universities and explained in handbooks. For example, the environmental toxicity of chemicals is established through relatively standard procedures, involving highly protocolled toxicity tests with model water organisms, assessments of expected distribution of chemicals through the environment, and ways of comparing expected exposure to expected toxicity to assess potential problems. For high-volume chemicals or chemicals of particular concern, additional assessments are required. In the context of regulatory toxicity testing, for the specific purpose of regulating which chemicals to allow onto the market, these protocols define what counts as trustworthy knowledge. Such procedures may be challenged: occasionally, researchers or actors may challenge whether these procedures are appropriate, 'scientific', fair, reasonable, or reliable (Chapman, 2007). However, most of the time such arrangements, resulting from long and fiercely debated negotiations, establish what can be considered sufficiently reliable knowledge in a given context.

From this perspective, the procedures and principles our societies have developed to certify knowledge are so much richer than relatively simple universal principles: standardised tests, accredited professionals, peer review, certified equipment, counter-expertise and second opinions, expert committee deliberations, public consultation, right-to-know procedures, transparency, and so forth. None of these procedures are a guarantee in themselves, and even their combined use may not always resolve things, but to replace them all with an appeal to 'the scientific method' or 'the norms of science' would sell short the rich resources our societies have developed to assess knowledge.

It is this wealth of practices and institutions that organise and adjudicate what counts as reliable knowledge in the context of environmental policy making that this book addresses. Now that we have removed our inherited tendency to fix this problem with a few simple rules of thumb (refutation, CUDOS, or insider knowledge), we can look at the actual wealth of practices and principles that have developed in the thick of environmental policy making, in the midst of controversies, fierce social debate, but also great successes. By now, there is a wealth of social science research into environmental policy making, the role of knowledge in policy, and the issues and tensions they give rise to. Our challenge is to find patterns in these debates and in the institutional arrangements that have developed. We must also articulate principles informed by both practice and research that can provide meaningful handles for environmental professionals to develop and use expert knowledge productively, modestly, and with integrity.

References

Abraham, C. (2004). *Possessing Genius: the Bizarre Odyssey of Einstein's Brain.* London: Icon.

Biagioli, M. (2002). From Book Censorship to Academic Peer Review. *Emergences*, 12 (1), 11–45.

Bohannon, J. (2013). Who's Afraid of Peer Review? *Science*, 342(6154), 60–65. doi:10.1126/science.342.6154.60

Bowker, G. (2000). Biodiversity Datadiversity. *Social Studies of Science*, 30(5), 643–683.

Callender, C. (2014). Philosophy of Science and Metaphysics. In S. French and J. Saatsi, eds., *The Bloomsbury Companion to the Philosophy of Science* (pp. 33–54). London: Bloomsbury Academic.

Callon, M. (1986). Some Elements of a Sociology of Translation: Domestication of the Scallops and the Fishermen of St. Brieuc Bay. In J. Law, ed., *Power, Action and Belief: A New Sociology of Knowledge?* (Vol. 32, pp. 196–229). London: Routledge and Kegan Paul.

Carson, R. (1962). *Silent Spring*. Boston: Houghton Mifflin.

Chapman, A. (2007). *Democratizing Technology: Risk, Responsibility and the Regulation of Chemicals*. London: Earthscan.

Cole, S. A. (1998). Witnessing Identification: Latent Fingerprinting Evidence and Expert Knowledge. *Social Studies of Science*, 28(5), 687–713.

Collins, H. M. (1975). The Seven Sexes: A Study in the Sociology of a Phenomenon, or the Replication of Experiments in Physics. *Sociology*, 9(2), 205–224.

Collins, H. M. (2014). Rejecting Knowledge Claims Inside and Outside Science. *Social Studies of Science*, 44(5), 722–735.

Collins, H. M., and Evans, R. (2002). The Third Wave of Science Studies: Studies of Expertise and Experience. *Social Studies of Science*, 32(2), 235–296.

Collins, H. M., and Evans, R. (2007). *Rethinking Expertise*. Chicago: University of Chicago Press.

Daston, L. J., and Galison, P. (2007). *Objectivity*. Cambridge, MA: MIT Press.

De Jonge, J. (2016). *Trust in Science in the Netherlands 2015*. Retrieved from Den Haag: www.rathenau.nl.

Desmond, A., and Moore, J. (2009). *Darwin's Sacred Cause: Race, Slavery and the Quest for Human Origins*. Harcourt: Houghton Mifflin.

Ellis, R., and Waterton, C. (2004). Environmental Citizenship in the Making: The Participation of Volunteer Naturalists in UK Biological Recording and Biodiversity Policy. *Science and Public Policy*, 31(2), 95–105.

Epstein, S. (1996). *Impure Science: AIDS, Activism, and the Politics of Knowledge*. Berkeley: University of California Press.

Fagan, Melinda B. (2010). Social Construction Revisited: Epistemology and Scientific Practice. *Philosophy of Science*, 77(1), 92–116. doi:10.1086/650210

Feyerabend, P. (1975). *Against Method: Outline of an Anarchistic Theory of Knowledge*. London: New Left Books.

Gauchat, G. (2011). The Cultural Authority of Science: Public Trust and Acceptance of Organized Science. *Public Understanding of Science*, 20(6), 751–770. doi:10.1177/0963662510365246

Geison, G. L. (1995). *The Private Science of Louis Pasteur*. Princeton: Princeton University Press.

Gieryn, T., Bevins, G. M., and Zehr, S. C. (1985). Professionalisation of American Scientists: Public Science in the Creation/evolution Trials. *American Sociological Review*, 50, 392–409.

Gould, S. J. (1981). *The Mismeasurement of Man*. New York: Norton.

Jardine, N., Secord, J. A., and Spary, E. C., eds. (1996). *Cultures of Natural History*. Cambridge: Cambridge University Press.

Jasanoff, S. (2003a). Breaking the Waves in Science Studies: Comment on
    H. M. Collins and Robert Evans, The Third Wave of Science Studies. *Social
    Studies of Science*, 33(3), 389–400.
Jasanoff, S. (2003b). Technologies of Humility: Citizen Participation in Governing
    Science. *Minerva*, 41(3), 223–244.
Kuhn, T. S. (1962/1970). *The Structure of Scientific Revolutions*. Chicago: University of
    Chicago Press.
Kwa, C. (2011). *Styles of Knowing*. Pittsburgh: University of Pittsburgh Press.
Latour, B. (1983). Give Me a Laboratory and I Will Raise the World. In K. D. Knorr-
    Cetina and M. Mulkay, eds., *Science Observed: Perspectives on the Social Study of
    Science*. Beverly Hills: SAGE Publications.
Latour, B. (1987). *Science in Action: How to Follow Scientists and Engineers through
    Society*. Cambridge: Harvard University Press.
Latour, B. (1993). *The Pasteurization of France*. Cambridge: Harvard University Press.
Latour, B. (2013a). Biography of an Inquiry: On a Book about Modes of Existence.
    Social Studies of Science, 43(2), 287–301.
Latour, B. (2013b). *An Inquiry Into Modes of Existence*. Cambridge: Harvard University
    Press.
Latour, B., and Woolgar, S. (1979). *Laboratory Life: The Construction of Scientific
    Facts*. Beverly Hills: SAGE Publications.
Law, J. (1987). Technology and Heterogeneous Engineering: The Case of the
    Portuguese Expansion. In W. Bijker, T. P. Hughes, and T. J. Pinch, eds., *The Social
    Construction of Technical Systems: New Directions in the Sociology and History of
    Technology* (pp. 111–134). Cambridge, MA: MIT Press.
Lawrence, A., and Turnhout, E. (2010). Personal Meaning in the Public Sphere:
    The Standardisation and Rationalisation of Biodiversity Data in the UK and the
    Netherlands. *Journal of Rural Studies*, 30, 1–8.
Leydesdorff, L., ed. (1980). *Philips en de wetenschap*. Amsterdam: SUA.
Lomborg, B. (2001). *The Skeptical Environmentalist*. Cambridge: Cambridge
    University Press.
Macfarlane, B., and Cheng, M. (2008). Communism, Universalism and
    Disinterestedness: Re-examining Contemporary Support among Academics for
    Merton's Scientific Norms. *Journal of Academic Ethics*, 6(1), 67–78. doi:10.1007/
    s10805-008-9055-y
McDonald, F. (2016). 8 Scientific Papers That Were Rejected Before Going on to Win
    a Nobel Prize. ScienceAlert (16 August). Retrieved from www.sciencealert.com
    /these-8-papers-were-rejected-before-going-on-to-win-the-nobel-prize
Merton, R. K. (1973 [1942]). The Normative Structure of Science. In R. K. Merton and
    N. W. Storer, eds., *The Sociology of Science: Theoretical and Empirical
    Investigations* (pp. 267–278). Chicago: University of Chicago Press.
Minnis, P. E. (2000). *Ethnobotany: A Reader*. Norman: University of Oklahoma Press.
Mitroff, I. I. (1974). Norms and Counter-Norms in a Select Group of the Apollo Moon
    Scientists: A Case Study of the Ambivalence of Scientists. *American Sociological
    Review*, 39(4), 579–595.
Mnookin, J. L. (2001). Scripting Expertise: The History of Handwriting Identification
    Evidence and the Judicial Construction of Reliability. *Virginia Law Review*, 87(8),
    1723–1845.

Oreskes, N., and Conway, E. M. (2010). Merchants of Doubt: How a Handful of Scientists Obscured the Truth on Issues from Tobacco Smoke to Global Warming: New York: Bloomsbury Press.

Pickstone, J. V. (2000). *Ways of Knowing: A New History of Science, Technology and Medicine*. Manchester: Manchester University Press.

Popper, K. (1934/1959). *The Logic of Scientific Discovery*. London: Hutchison.

Popper, K. (1942/1966). *The Open Society and its Enemies* (revised fifth edition edn.). London: Routledge and Kegan Paul.

Radder, H. (2010). Mertonian Values, Scientific Norms, and the Commodification of Academic Research. In H. Radder, ed., *The Commodification of Academic Research* (pp. 231–258). Pittsburgh: University of Pittsburgh Press.

Rip, A. (2003). Constructing Expertise: In a Third Wave of Science Studies? *Social Studies of Science*, 33(3), 419–434.

Shackley, S., and Wynne, B. (1996). Representing Uncertainty in Global Climate Change Science Policy: Boundary-Ordering Devices and Authority. *Science, Technology, and Human Values*, 21(3), 275–302.

Solomon, S. M., and Hackett, E. J. (1996). Setting Boundaries between Science and Law: Lessons from Daubert v. Merrell Dow Pharmaceuticals, Inc. *Science, Technology, and Human Values*, 21(2), 131–156.

Van Zwanenberg, P., and Millstone, E. (2000). Beyond Skeptical Relativism: Evaluating the Social Constructions of Expert Risk Assessments. *Science, Technology & Human Values*, 25(3), 259–282.

Verran, H. (2001). *Science and an African Logic*. Chicago: University of Chicago Press.

Wittgenstein, L. (1953). *Philosophical Investigations*. Oxford: Blackwell.

Wynne, B. (2003). Seasick on the Third Wave? Subverting the Hegemony of Propositionalism: Response to Collins & Evans (2002). *Social Studies of Science*, 33(3), 401–417.

Ziman, J. (1994). *Prometheus Bound: Science in a Dynamic Steady State*. Cambridge: Cambridge University Press.

3

Frames: Beyond Facts Versus Values

WILLEM HALFFMAN

3.1 What Are Frames?

Have a look at Figure 3.1. What is this?

This question has many factually correct answers. This is a landscape. It is a waterway. It is a stereotypical image of Holland's countryside. It is a tourist site. It is an example of reclaimed land. It is a picture of monumental windmills. All of these, and quite a few more, are arguably correct answers. All of them *portray* this image in a certain way, focusing on one aspect or another. Calling this a *landscape* gives a particular meaning to this image as related to other

Figure 3.1 Mill Network at Kinderdijk-Elshout, Netherlands. © Douglas Allen Deacon / Getty Images.

landscapes: mountains, or deserts, for example. Describing it as a representation of a monument puts the image in a family consisting of churches, medieval castles, and the Eiffel Tower.

This process is called framing. Frames identify and give meaning to a situation by defining what it is, what facts about it are the most relevant, and what other situations it is related to. Frames tell us *what kind* of situation is in place and how we should interpret it.

In common parlance, 'framing' can have a rather negative connotation: for example, setting someone up (framing someone for a crime), or putting a misleading 'spin' on a story. This sometimes leads to confusion, since pointing to framing can be interpreted as an accusation of falsehood. This is not our intention at all. We will use the term here to refer specifically to how people interpret and understand situations, as in the example above, based on the observation by hermeneutic philosophers that the environment comes to us always already interpreted (Drenthen, 2017).

It is important to point out that the difficulty in establishing what the wind-mill picture *really is* is not a problem of lack of facts. An outrageous amount of facts are known about this place, about its history, water levels and water quality, its biodiversity, its economy, pollution levels, hydraulics, ... You could even claim that the difficulty of determining what this *is* is not so much a problem of *lack* of facts, but of an *over-abundance* of facts. If you want to know 'the facts' about this situation, the only manageable reply would be 'Which ones do you need?'. Frames focus our attention on which of the available facts matter. If you are a photographer or graphic designer, the quality of the colours may be the most important aspect of this photograph. To a lawyer or publisher, the ownership of its copyrights is the most important fact. Frames establish what is relevant.

It is even possible to claim that the image is a misrepresentation of this particular situation. For example, there are no people in this image, which can be criticised as a skewed representation of what is actually a living community. Or maybe the image fails to represent problematic soil pollution, or the modern houses and highways that exist beyond the scene captured in the image, suggesting some pristine idyll of the past where there is none. You could claim that the picture is *biased* because it is mis-framed. There may be very good arguments for claiming that a particular frame is appropriate, supported by very solid facts, but it is impossible to resolve what is the *right* frame by using *only* facts. At the same time, some attempts to frame this situation would be factually ludicrous: this is not Mount Everest or a picture of my dog. (In fact, I do not even have a dog.) Nevertheless, facts alone do not allow, and do not force us to express what this is and is not.

Good reasons to claim how this situation should be framed could be found in what should be done or what should be valued. In this case, the old windmills are valuable historic buildings: they constitute Dutch cultural heritage, worthy of protection. Therefore, there are good reasons to frame this as a picture of monuments. Alternatively, you could argue that this place should be seen as a valuable natural environment, to be protected from urbanisation. Now the most relevant facts are not the age or rarity of the windmills, but the local biodiversity, the uniqueness of the landscape, or the endangered status of the bats living in the windmills' rafters. You could also claim, with good reasons, that this is a situation of water management, in which the old windmills are no longer able to pump enough water out of this sinking, reclaimed, lowland peat. In this last frame, the windmills are an outdated technology that needs to be replaced if human activities such as agriculture are to remain possible here. Framing a situation in a particular way is argued with facts, but in every case, different facts are considered the most relevant based on what is valued.

If actors quarrel about how a situation should be interpreted, then such 'frame conflicts' can be very hard to resolve. It is hard to say which of the above interpretations is *right*, especially on factual grounds. Crucial facts in one frame may be meaningless in another, or could even mean the opposite: the age of the windmills can be seen as an argument both in favour of and against their protection. Windmill enthusiasts might invoke the age of the windmills as crucial evidence in defence of their protection, but from the perspective of modern water management, the age of the windmills highlights the need to replace them with more modern technology. The smart policy maker may understand this frame conflict and find clever ways to reframe the issue: a 'culture landscape' might combine ecological and heritage-based meanings. Reflection on frames can sometimes lead to a breakthrough in controversies, with new interpretations that accommodate a wider set of values and facts.

In environmental policy, frame conflicts are rife. Be it fisheries, pollution, climate change, regional planning, water management, or nature conservation, debates often extend beyond measurements, experimental results, or model predictions to what such factual information means and how it is relevant. Especially in controversial environmental issues, actors may completely dis-agree about how issues should be framed. For example, the long-standing controversy over genetically modified organisms (GMOs) in agriculture is partly about whether GMOs should be understood as just another breeding technique, or as a radical technological break with the past. In the first case, GMOs are framed as similar to the crossbreeding that farmers have been doing since the dawn of agriculture; in the second case, GMOs are framed as a highly technological disruption of that tradition. Similar tensions arise when proponents see GMOs as just a farming technique without significant health risks, while opponents see it as an injustice in power relations between the global North and South. No amount of insisting on lack of health risks will overturn opposition on the grounds of political arguments about the rights of farmers. It is therefore essential that environmental scientists, policy makers, and other actors are able to recognise and identify framing issues. Clarification of diverging frames may not always allow issues to be resolved, but it at least helps to articulate what any disagreement is *about*.

3.2 Framing and the Sciences

In our experience, framing is a strange and unwieldy phenomenon to natural scientists. They often find its ambiguity uncomfortable or even antithetical to science. ('Vague', my students say, which is one of the worst sins, in their book.) This unease is understandable. Part of the strength of natural science is its capacity to neutralise framing issues as soon as possible by defining the problem and thereby stabilising the frame. For the molecular biologist, the cell is a bag of molecules. If it is ambiguous which molecules are relevant, then this should be articulated. If the boundaries of the cell system are ambiguous, then it should be defined. Natural scientists are trained to remove ambiguity by defining and demarcating situations until they are clear and manageable, so they can go to work at what they do best: calculating in a well-defined system, measuring in a well-controlled experiment, and reasoning with unambiguous theories. Interpretation is to be eliminated and replaced by unquestionable fact, in which the 'subjective interpretation' of the scientist no longer plays a role. However, in society there are no controlled conditions, no authority strong enough to impose a definition, especially not in a democracy. Imposing one frame so we can proceed with measuring and researching may make it easier for researchers to work with a well-demarcated problem, but this may mean that some actors see their valued facts disappear from the equation. What is more, scientists warn us that a degree of scepticism about scientific assumptions is healthy, because facts and theories are not always as unquestionable, unambiguous, and uninterpreted as the most hard-nosed defenders of 'solid scientific truths' might suggest. The sciences periodically question the way they understand cells, ecosystems, or the universe.

The opposition between facts and values is deeply rooted in Western thinking. It draws a strict line between registering what is going on in the world ('what are the facts?') and what should be happening (values, goals, preferences): that is, what *is* and what *should be*. This provides a simple but highly problematic principle for the organisation of knowledge and decision making: scientists provide the facts, and then citizens, politicians, and all interested others express their values and decide what to do. If scientists can describe the facts about geo-engineering, then we citizens can decide whether we want it. This principle seems to allow scientists to 'stick to the facts' and avoid getting entangled in politicking.

However, frames complicate the matter considerably. Depending on how we define an environmental problem, certain facts become relevant while others disappear from view. If you define air pollution as a human health issue, then facts about the effects of air pollution on public monuments or the green environment disappear from view – no matter how hard they are. To claim that scientists 'deal with the facts' and political actors 'deal with the values' fails to explain how we come to agree on a meaningful problem definition, which neither can do on their own. If there are facts and values *and* *frames*, then things get more complicated than a simple two-way division of labour.

If you are a social scientist, then you might be more familiar with framing – especially if you are a social scientist in the tradition of qualitative research. The importance of framing in daily interactions between people was described by American sociologist Erving Goffman, an extremely perceptive observer of social life. He described how people behave differently according to how they interpret a situation, and how we can find patterns in such interpretations. For example, Goffman described the contrast in the behaviour of waiters, who, while friendly and servile in the restaurant, cuss the very same customers in the kitchen. This does not mean that the waiter is a hypocrite or a liar, but merely that the codes of restaurant and kitchen interaction are different, requiring participants to understand where they are in order to know how to behave. Goffman described such contrasts as 'front stage' and 'back stage' behaviour, and showed how such contrasting frames occur throughout social life (Goffman, 1959).

Goffman worked in the tradition of symbolic interactionism in the social sciences, which posits that people do not act in response to their material or social environment directly, but via the lens of how they interpret their environment. This can be applied to an environmental example: people do not respond to environmental degradation as such, but to how they interpret this degradation. If they perceive it as an inevitable side effect of modernisation, the response will be quite different than if they perceive it as a problem caused by their personal actions. If you want people to take action, it is not only important to understand the state of the environment or the economy, but also how people interpret these conditions.

Frames are also an important concept in the study of public policy, a field of social sciences in its own right, in a particular approach called Interpretative Policy Analysis (Hajer and Wagenaar, 2003; Metze, 2014; Pijnappel, 2016; Yanow, 1996). This approach focuses on how policies and the knowledge they are based on acquire meaning, and on how interpretation drives choices. For example, if applied to flood control policies, this approach can clarify how policy makers and scientists alike may 'get stuck' in a particular tactic of higher levees and large-scale flood defence works. Such a frame, supported by powerful institutional players in Dutch water management, pushed aside alternative knowledge and solutions, until a crisis offered opportunities for ecological water management and more local and participatory solutions (Van den Brink, 2010). In contrast, 'rationalist' approaches to policy will focus on measurement – for example, comparing the costs and benefits of policy alternatives, within a fixed frame: should the levee be rock or concrete? Similar to the engineers, economists can start their calculations for such assessments as soon as the contours of the problem are fixed. In contrast, interpretive analysis will highlight shifting problem definitions, such as disagreement about what constitutes a cost or a benefit (and for whom).

Especially in complex, open-ended problems, as is often the case in environmental policy, awareness of frames may help to pinpoint problematic assumptions and hidden disagreements, or may simply help to generate out-of-the-box solutions. If you can reflect on the framing of a particular issue, then perhaps an alternative framing of the issue becomes possible. We will return to such 'frame reflection' (Schön and Rein, 1994) and other practical uses of frames at the end of the chapter. However, in order to reflect on frames, you first have to be able to recognise them. While scholars such as those from Interpretative Policy Analysis have developed schemes and concepts for quite detailed 'frame analysis', a few pointers can get you started on recognising frames and making them explicit.

3.3 How to Identify Frames in Language

In the introductory example to this chapter, we used a photograph to illustrate frames. However, we can also identify frames in language: in the use of particular words to characterise a problem or a situation, in the use of metaphors, or in definitions. Frames can even be embedded in material instruments, buildings, or infrastructure. We will focus on frames in *language* first, and particularly in the documents that figure in environmental expertise and policy: policy plans, advisory reports, and scientific articles.

Generally, the best clues for frames in such documents can be found in the introductions of texts. It is where problems get defined, things get named and characterised, and approaches and perspectives get selected. Take the following sentences from an article on the pollution of the river Danube:

> the actual effect of chemical mixtures on aquatic organisms in the field is still largely unknown. Studies evaluating the influence of single chemicals or chemical mixtures on aquatic communities are not able to explain the majority of the variation observed.
>
> *(Rico et al., 2016)*

In these sentences, the topic of this article has been identified as an *ecological* question: the effect of chemicals on aquatic *communities* in the field. The topic is not framed as a question of the toxic effect of one chemical on one organism (as toxicologists tend to approach this), but on a community of organisms consisting of different species. The problem is defined as how to assess the effect of mixtures of chemicals in the field. This frame contrasts with the dominant frame in environmental risk assessment of chemicals, which generally assesses risk on a substance-by-substance basis on individual species (although there are exceptions). The quoted statements delineate what will (and, implicitly, will not) be studied, and why this approach is meaningful: the variation of ecological effects can allegedly not be explained from effects on single species. In this problem definition, we find the first indications of how pollution is framed in this text.

In order to identify frames in such texts, you can follow three simple and pragmatic strategies: identify what is included and excluded, identify metaphors, or identify storylines. There are some technical advantages and disadvantages to each of these approaches and you might develop a preference for one or the other, but essentially all of these approaches offer different paths to the same goal. For all three, you will need either knowledge about the context or texts with different approaches to the same topic in order to bring frames into contrast.

3.3.1 Inclusion and Exclusion

The first approach is to identify what is and is not included in the text. The Danube example article is about pollution in the Danube, not in its tributaries, and not the resulting pollution in the Black Sea or of the ground water. In addition, it is limited to benthic communities (organisms living in or near the sediment) of invertebrates (excluding other water organisms such as plants or fish), measured at 66 locations along the length of the river (and not other potential locations). The river is thus framed as a series of points with bottom-dwelling habitats, followed by an analysis of how these are influenced by environmental 'stressors': changes in water flow, alterations in river banks, and chemical pollution (see Figure 3.2, reproduced from the article). The assumption is that meaningful claims can be made about the ecology and 'stressors' of the entire river based on these points. For this study, the entire length of the river was considered relevant. (A study assessing national policies might have limited the river to just the Hungarian stretch. Another approach could have been to study the entire river basin, not just the Danube itself.)

Geographic delimitations are one form of *scale frames* (Van Lieshout et al., 2011), here defining the territory a study or policy applies to. Another form of scale-frame is delimitation in time: to what period does the study apply? Time frames are particularly important in debates over risks, for example in whether very long-term effects of chemicals, radiation, or waste can or should be included in risk assessments. For example, in the case of nuclear energy, very low probabilities of malfunction may produce significant undesired effects if considered over a sufficiently long period of time. Clearly, scale frames refer to responsibilities and problem ownership. Whether we should include long-term effects in assessments, or cross-boundary pollution, depends on how we understand our responsibility for adverse effects to future generations, or countries' responsibilities to their neighbours.

To be sure: our first intent is not to judge whether these inclusions and exclusions are *correct* or *opportune*, or whether the assumptions can be *justified*. (Certainly, the expert reviewers of the journal in which the article was published thought this demarcation was entirely justified, since the article was published.) We merely want to identify what is and what is not part of the study and make frames explicit and available for reflection.

Obviously, the list of what is *not* part of our example study is infinite: the phase of the moon is not considered, nor songbirds along the Danube, nor Superman. Pointing out that all those things are not in the article would just be silly. However, some other exclusions could arguably play a role in the pollution of the Danube. This is where a comparison of frames in different texts becomes interesting. How do Hungarian policy documents frame the pollution problem in the Danube, in contrast to these international studies? Would policy documents also see this as a problem affecting benthic communities, or is human health the primary concern? How do water sanitation engineers frame the water problem? In public debate or policy discussion, such different understandings of where the problem begins and where it ends may underlie disagreements that are phrased only as competing statements about facts.

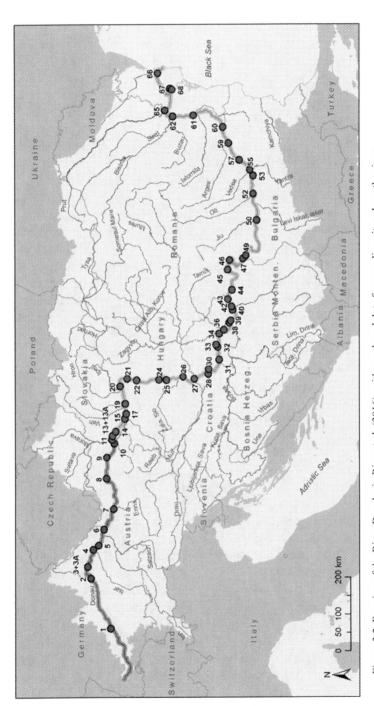

Figure 3.2 Framing of the River Danube in Rico et al. (2016), with numbered dots for sampling sites along the river.

3.3.2 Metaphors and Comparisons

The second strategy is to identify metaphors and comparisons in the text that characterise the topic in a particular way. The Danube article investigates 'abiotic stress factors' such as chemical pollution, dams, and riverbank alterations on invertebrate communities (Rico et al., 2016, 1373). By using 'stress' and investigating 'stressors', these communities are implicitly compared to an *organism* under stress – just like people can be under stress from working too hard, smoking, or disease. Similarly, engineers talk about a bridge being 'under stress'. The article also talks of *pressure* on these communities, referring to a physical process such as hydraulic pressure. Hence, the language of 'stress' and 'pressure' seems to indicate that the invertebrate community is understood as a system (a functional unit with a clearly defined boundary) that is either mechanical or organic in its nature. These communities are analysed *as if* they were organisms exposed to unhealthy influences. Alternative frames could have been 'disturbances' (suggesting the benthic communities were at peace, in some form of equilibrium), or 'effect' (not suggesting any form of natural state). Variation in such terms and their connotations sometimes leads to extensive debates about which language is more appropriate. In the case of ecology, they can refer to deeply held convictions about stability or equilibrium in nature, or even about what is to be considered 'nature' or 'natural' in the first place.

By merging different abiotic changes in the river into the larger category of 'stressor', the article also merges very different phenomena into one category to make them comparable: such diverse things as dams, chemicals, and channelisation are put together into one 'stressor' category, based on indicators that make them *similar*. In many ways a dam and a pesticide are two entirely different things, but from the perspective of environmental stress, they are actually quite like each other: they are similar *in some respects*. Hence, categorisation can also be a fruitful route to study comparisons or metaphor in framing: what larger group of phenomena does something belong to? The title of the article does something similar: 'Relative Influence of Chemical and Non-chemical Stressors on Invertebrate Communities: A Case Study in the Danube River'. The Danube invertebrate communities are a *case*, one member of the larger collection of riverine invertebrate communities. Just as the monument frame for the windmill related our windmills to churches and castles, the specific selection of invertebrate communities implicitly relates the Danube to the ecology of (aquatic) invertebrates all over the planet.

As a natural scientist, you may find this puzzling. Most scientists will aim to *remove* metaphor from scientific publications: metaphor is a matter for poetry, too imprecise for scientific work. If I compare you to a summer's day, then the similarities are quite a stretch and allow for too much interpretation by the reader – arguably a great quality of poetry, but not of exact science. Hence, scientists will try to replace metaphor with precise definitions that remove interpretive flexibility as much as possible. To be scientific, I will compare your underarm body temperature to that of a July noon average, as measured by a standard weather station, rather than an imprecise 'summer's day.' Nevertheless, connotations and assumptions remain, such as about the fundamental nature of the phenomenon at hand: I will compare temperature, not smell. In the case of Danube pollution, the notion of 'stressor' is what makes hugely different phenomena comparable. By now, the concept of 'stressor' has received a precise definition in environmental science, but even then the basic assumptions of what was once a more poetic metaphor remain in place.

In fact, scientific writing can be surprisingly metaphorical. Darwin's *On the Origin of Species* is full of metaphors, comparing the relation between competing species with competing firms in 'the economy of nature', or the struggle to survive in nineteenth-century capitalism (Young, 1985). Darwin apologised for the use of such metaphors and tried to replace a few with more neutral expressions in later editions, but some critics continue to point out that Darwin focuses on competition between individuals at the expense of the many examples of fruitful cooperation in nature. The metaphorical language may have been polished, but some of its assumptions remain.

Even modern scientific texts can contain quite explicit metaphors. In recent years, there has been a fierce debate about what constitutes appropriate terminology in 'invasion biology', which studies non-native species and their effects. The very term 'native' is already problematic, as it typically refers to national borders of little ecological relevance. Some biologists have even objected to the xenophobic connotations in invasive biology, or to terms such as 'invasional meltdown' or 'bio-pollution' (portraying invasive species as similar to chemical pollution). Such metaphors may be used with especially strong effect when explaining scientific findings to larger audiences, but can be equally indicative of frames used in research itself (Verbrugge et al., 2016).

3.3.3 Storylines

The third technique to identify frames is to look at the narrative: what is the story being told? Who are the protagonists of this story and what happened to them? The Danube article can be read as the story of five scientists who collected samples of invertebrates from the bottom of the river and compared their samples to the river's physical characteristics. In customary scientific parlance, their role as active protagonists is carefully removed with passive tense: the data 'were gathered', sampling sites 'were selected' and 'were classified'. The heroes of the story remain unsung.

However, on another level, the heroes and the victims of the story are much more explicit: the invertebrate communities of the Danube (the victims) are threatened by stressors (the villains of this story) and the article will point out the worst of these stressors, so that governments (the heroes) can intervene with specific 'risk management options'. By turning the story into a fairy tale with heroes, victims, and villains, we can clarify what the authors see as their role in the world: to provide the authorities with knowledge that will allow better protection of the environment. Here, too, there are some interesting assumptions: that governments care about benthic invertebrate communities or the riverine ecology they belong to, or that governments might *act* on the basis of such specific information if it is properly delivered by the scientists. If this fairy tale approach sounds facetious, it is only because such assumptions remain implicit in these kinds of texts and we need a procedure to magnify and articulate them – these assumptions may be perfectly justified.

An analysis via storylines allows us to see how *agency* is framed: which actors are seen as strong or weak, which are expected to act, which are expected to help (and how). Here, too, contrasts with other storylines could help. For example, environmental activists could portray governments as slow and inactive, or see economic relations as the ultimate culprits of environmental problems (rather than a lack of 'risk management options'). The scientists in the Danube article are also competing with other sources of advice, involving risk assessments based on individual chemical substances and species. Their claim is that their approach allows for more effective protection of the river. Contrasting storylines of different actors bring the specific nature of the frame into sharper focus.

The 'heroes, villains, and victims' list is but one very simple scheme to identify storylines. A similar device to identify frames is to look at how issues are identified, how issues are categorised, how responsibilities are attributed, and how causal connection are identified (Pijnappel, 2016; Scholten, 2008). In the Danube example, the issue is *identified* as an issue of ecological river pollution (not human health or fish toxicology). The issue is *categorised* as a case of an entire river, divided into specific measuring locations, with different forms of disturbance grouped together as 'stressors' (as opposed to a river basin, with point-source pollution discharges, or other ways of organising knowledge about this river). *Responsibilities* are distributed between the scientists who measure and the governments along this river who should choose. Last, the *causal connections* are made between specific causes (pollution, physical changes) and specific effects (in benthic communities). Different or additional causal connections could have been included, such as secondary toxic effects on fish. These simple techniques can help make the narrative of an environmental policy text or research article explicit in a more systematic way.

Storylines can provide the basis for strong coalitions among actors. If the scientists, governments, or other actors in the policy field share the same idea about how the world is to be made a better place, about how their roles complement each other, and about how the world is to be understood, then this could facilitate cooperation. For example: 'we experts will predict what will happen in the mechanism of the economy after policy makers take specific measures, so policy makers can choose what to do based on their preferred outcomes'. In this storyline, the world is like a machine, with 'levers' that policy makers can pull, which will have outcomes that can be predicted by economists and their models (Halffman and Hoppe, 2005; Kwa, 1994). Such shared understandings of the world and mutual division of labour can form powerful 'discourse coalitions' that can dominate thinking in a policy field over long periods, even if the shared understanding turns out to be ambiguous (Hajer, 1995; Pijnappel, 2016; Stone, 1997).

3.4 Institutional Frames

When people act upon the world by interpreting it through a particular frame, they will organise the world according to that frame. To frame the river Danube as a series of benthic communities under stress, you need particular sampling instruments, the organisation of a monitoring network to gather data, and resources to analyse that data. If this approach to river pollution assessment was to prove useful for water authorities in the basin, they might make it part of long-lasting monitoring schemes, or even embed them in regulatory requirements. This means frames are not only present in the way articles or policy documents *talk* about the environment, but also in material objects such as measuring equipment, or research and policy practices, or even longer-term patterns organised into such practices or *institutions*. Hence, frames can be embedded in devices, interaction between people, or organisations, whether or not they are made explicit in documents.

For example, if we study river pollution as a problem of sewage pipes discharging industrial or residential effluent, we would want to assess particular parameters: we would want to know how much water flows through a river to calculate how the effluent is mixed, and how much effluent flows into it. For this, we would need to put in place a measurement network to record flow data.

If we want to build environmental policy on this, then we might develop discharge quality standards to make sure we can set fair limits for all dischargers. We would have to develop limits for specific chemicals in effluent or generic indicators of the effluent's toxicity. To make sure discharges meet such limits, dischargers would have to build water treatment plants. This is the kind of river water quality approach that was developed in most industrialised countries in the 1970s. Its key features were to address river water pollution at the point of discharge with filters and other treatment technologies. By the 1980s, these became known as 'end-of-pipe solutions': a particular way to frame river water pollution. By then, the end-of-pipe frame had been built into policies, measuring devices, standards, and physical infrastructures. In the decades that followed, environmental policies would develop other frames, relying more on prevention and on integration of water and other environmental compartments, for improving river water quality.

This 'delegation' from language to less explicit carriers of frames complicates things. In order to be able to reflect on frames, you have to be able to make them explicit by articulating them in words, images, or other symbolic representations. If frames are implicit and deeply embedded in things and routines, they may not be easily identified and will require extra work of interpretation by us. We can no longer rely on the language and words already provided in texts; we need to interpret the implicit assumptions of measuring devices, waste management practices, or equipment. That is what the environmental scientists and activists of the 1980s did: they pointed out the limitations of the existing ways of dealing with river water pollution by highlighting the particularity of its frame.

Hence, we can trace how frames become embedded in organisations, procedures, and routines, or even in material objects. If we act as if the world is a machine, then we will gradually start to organise the world *like* a machine. If we assume that human beings act as rational economic actors, then we will start to organise the world as if it consists of rational actors, along with economic models and indicators to sustain this frame. Frames then become more than mere text – they become an organising principle for shaping the world around us.

In this sense, you can claim that institutions such as the economy, or environmental policy making, have particular frames: they 'perceive' and 'process' the world in a particular way. In a famous example about different frames in the wine industry, anthropologist Mary Douglas' text *How Institutions Think* showed how the French prioritise region in the categorisation of wines, while the rest of the world gives more attention to the kind of grape. For the French, wines that come from the Bordeaux region belong together, even if they contain different varieties of grape. The territorial origin (the lay of the land, soil type, and climatic conditions or '*terroir*') is considered the dominant influence on French wine's character. This difference of categorisation is not just a matter of superficial convention, but a difference that is embedded in how wine is labelled, valued, and traded, and in how the wine industry is organised (Douglas, 1986).

Similarly, we could claim that the dominant Western way of framing nature in the industrial era, in which nature was seen as an infinite resource for human production, has gradually given way to a different understanding of a more finite and fragile world. This frame is not only present in our language and thinking about the environment, but now also in waste-water treatment plants, environmental campaigns urging us to behave like responsible citizens, and recycling bins and tariffs that nudge us into behaving responsibly.

This raises an important question: where are these frames located? Are they somehow 'underneath' our environmental policies, driving us in a particular direction, urging us to dig them up and make them visible? Or are these frames merely ways in which we interpret the world, by imposing our interpretation on top of the world? Do we discover frames in our approaches, or do we invent frames to identify the limitations and assumptions of our approaches?

3.5 The Nature of Frames

3.5.1 Frame or Framing?

By now, it should be clear that identifying frames is not an exact and purely deductive process. Identifying and reflecting on frames requires *interpretation* on our behalf. There can be good arguments to interpret a text in a certain way, but someone else might come up with another interpretation, identifying another frame, such as by choosing different contrasting texts. Contrasting the Danube article with a water sanitation text or with a physiological study of toxicity to individual water organisms will highlight different frames. You could therefore claim that the frame is not 'in' the text, but in the way we interpret the text and contrast it with other texts. This tricky problem is actually quite consistent with the central claim: if the world is always already interpreted as we deal with it, then so are the environmental sciences and policies in which we are trying to identify frames. In that sense, identifying a frame is more precisely described as our own reconstruction of other actors' frames, which will of necessity always involve some reinterpretation (Pijnappel, 2016, 42; Van den Brink, 2010, 42–44).

Based on this argument, there are two fundamentally different approaches to studying frames. The first is to assume that frames are a 'thing', hidden in texts or practices to be revealed by us. The second approach insists that frames arise in interaction, in contrasting texts, or in a conversation between people: as they talk, they gradually come to define their conversation and act accordingly. The school that insists on the interactional nature of the process will therefore prefer the term 'framing', which stresses the process. The term 'frame' refers more to an underlying structure. One of the major arguments in favour of framing as a process is that it shows that frames are flexible and negotiable in human interaction, avoiding the suggestion that we are somehow victims of the invisible powers of frames. The argument in favour of an analysis of frames as structures is that it may clarify assumptions in our thinking, so we can overcome them. From a more fundamental point of view, frame and framing can be seen as two complementary moments in how we interpret the world and shape it based on our interpretations.

3.5.2 What Frames Do

Frames help people identify what matters in the world. They focus attention and help make sense of complex situations. They indicate what is valuable and how situations should be understood, so we can work on gathering facts and looking for solutions. For example, a forest is an incredibly complex environment, but if you see a forest *as* today's location for a stiff walk, then you know you need to pay attention to paths, road signs, and other points of orientation. If you are a logging company, you most likely identify different features of the forest as pertinent, such as the volume of economically harvestable wood. Rather than wallow in the overwhelming and stifling complexity of the world, frames tell us where to start and what to ignore, at least for the time being. Frames reduce complexity – perhaps not in the world out there, but at least in the way we deal with the world.

By selecting what is important and which facts are relevant, frames also point to which forms of knowledge are relevant. For the walker, a clock, hiking maps, a compass (or GPS), and a walking guide may provide relevant forms of knowledge, while the forester's special knowledge of how to measure the volume of timber will not be so relevant. For nature conservation, precise knowledge of biodiversity and ecological knowledge about how to create suitable conditions for vulnerable species will be of prime importance. We can also turn this around: from the perspective of specific forms of knowledge, the forest looks like a different place. The walking guide, forestry science, and forest conservation ecology will all portray the forest differently.

They will collect and highlight different facts about the forest. Hence, different forms of knowledge, as practised by different professions, disciplines, and even everyday users, will frame situations in the environment differently.

Consequently, a different frame, and the knowledge that comes with it, will also suggest different solutions or ways forward. To the walkers, the forest may need additional paths, leading to more places of outstanding beauty, to the most elusive plants and animals, or to the most adventurous locations. To the conservation ecologist, the solution to the forest's problem might be to close it off to walkers completely, or to allow a minimum number of walkers to sustain some public support of conservation. From the perspective of timber management, harvesting wood economically may require a solution of standardised forest for production purposes (Scott, 1998). Especially when proposed measures become concrete, such as with extra roads, or more fences, or felling, such different understandings of what is relevant about the complexity of the forest can lead to pointed disagreements.

3.5.3 Frame Dynamics

By now, it may sound as if we are victims to our frames, as if frames are irremovable lenses. Or it could seem that we have no alternative but to 'agree to disagree': you see the forest your way, I see the forest my way, and that's that. However, the situation is not that dire. Frames are not static and unchangeable devices to interpret the world, but rather offer some degree of flexibility and room for development. For example, as a walker using the forest, we may develop a greater appreciation of the ecology of the forest if the conservationists can manage to make some of its valuable species more visible. Logging may proceed in ways that can be reconciled to some degree with nature conservation, if appropriate techniques and strategies are adopted. Even though important and sometimes painful differences between forest uses remain, forest management has changed considerably over the last century, seeking new concepts and practices that accommodate uses and insights from different frames.

Adapting and changing frames intentionally can be seen as a particular form of *learning*. As a conservationist, we can learn more and more about biodiversity, adding more facts and causal relations that fit with the way we perceive the forest. We call this 'primary learning'. However, if we identify the limitations of our perspective of seeing the forest as biodiversity, perhaps in contrast to other perspectives, then we might find ways to improve our dealing with the forest by changing how we see the forest. For example, if the conservationist includes people among the interesting biodiversity of the forest, then new management options may become available. This is called 'secondary learning' – that is, learning about the conditions of your knowledge and improving your perspective, rather than just adding more facts to it.

Frame changes or 'secondary learning' may happen in different forms. One frame might be replaced by the next. For example, there was an old frame of seeing indigenous people as part of 'nature', as opposed to the 'culture' of Western colonial powers (Jardine et al., 1996). Nowadays, this framing of indigenous life is no longer accepted. Two opposing frames might integrate to form a more encompassing understanding of the world. For example, we might attempt to develop an environmental policy that keeps the achievements of the end-of-pipe technologies such as water treatment plants, but see water treatment plants as part of larger cycles in human ecology. Or, two frames might not quite integrate into one, but co-exist in a mutual accommodation, perhaps linked through partially shared elements such as shared narratives or devices that take on different meanings in the different frames. Or, two frames might develop in opposition to each other, articulating better arguments as they confront each other – or variations on such patterns.

In order for such changes to occur, older frames, which may be embedded in organisational structures and practices, may have to be 'dislocated' (Van den Brink, 2010, 29–31). For example, in nature conservation, some ecologists have increasingly come to appreciate the role of people in shaping nature throughout history, dislocating the frame of 'wild nature without people' that was fundamental to the creation of wildlife parks. Hence coppicing, selective felling, or even some degree of livestock grazing may become thinkable as viable conservation instruments in forests. At the same time, the 'wild nature without people' frame is still fiercely defended in some wildlife conservation practices that aim to minimise any human interference. In these examples, different nature conservation frames confront each other. If dislocation is successful, it may lead to an accommodation between conservation practices, or even integration within a more encompassing understanding.

You might think that integration of frames is an obviously superior form of learning. However, frame integration may come at a price: in order to really integrate different frames, they may have to be made entirely compatible, perhaps at the expense of what made the frames unique. The 'wild nature without people' and a 'wild nature with people' frames may find some accommodation, but there will always be some degree of compromise, not only in how nature is dealt with, but also in how it is known. The disadvantages of integration over maintaining diversity are elaborated in the case of climate policy provided (see Case A: Framing Climate Change) and discussed in more detail later in the book (see Chapter 7).

3.6 The Relevance of Frame Reflection

Policy scientists Schön and Rein have argued that professionals should be aware of how their understanding of problems and solutions has its own particular frame. An ecologist is more likely to understand a problem of forest management as an ecosystem problem, while an agricultural scientist might give more attention to harvesting and returns. Schön and Rein advocated that professionals should be able to reflect on their framing bias, and be aware of its limitations and differences from other approaches. With this 'frame reflection', professionals should therefore have more than just expert knowledge of techniques or methods; they should also be aware of assumptions and their particular bias, in order to come to more fundamental improvements in their knowledge (Schön and Rein, 1994).

Schön and Rein saw frame reflection as an essential exercise to break out of 'intractable policy controversies' – that is, the kind of controversies that drag on for years, in which policy actors do not seem to agree on what facts and values are at stake, nor even on what the problem actually is. These problems are sometimes also called 'wicked problems' (Hisschemöller and Hoppe, 1995–1996; Hoppe, 2010; Kreuter et al., 2004), because they seem to defy attempts to resolve them with traditional policy-making tools. The debate over the desirability of genetic modification of agricultural crops (mentioned earlier) is one such controversy, just as the debate over nuclear energy was (and in some countries still is). By pointing out different assumptions about what does and does not belong to the problem, which facts are relevant for whom, and how different actors reason from different assumptions, frame analysis makes points of disagreement explicit and opens them up for debate.

From the position of experts and civil servants working in environmental policy, frames may play a role as one particular form of uncertainty. Whereas limitations of measurement equipment or available data may be a source of uncertainty over data precision or knowledge uncertainty, framing uncertainty specifically concerns the question of whether a problem has been adequately defined. Here, 'adequate' may refer to whether pertinent facts have been included, but also to whether contesting problem perceptions have been accounted for (Halffman and Ragas, 2018, forthcoming).

Frame analysis can also be performed from a more critical perspective, in order to expose or challenge assumptions and preconceptions. Showing how an environmental problem is approached in a particular way can serve to propose alternative framings, which could offer new or better possibilities to solve environmental problems or build new coalitions for environmental action. For example, if the problem of micro-plastic pollution in rivers is framed as just a problem of river pollution, then marine pollution effects remain out of scope. However, if the marine environment is included, then the potential contamination of fish is included in the analysis; also, fisheries may become a powerful political interest that could help raise concern for this problem (Halffman and Ragas, 2016). Hence, reframing issues can bring powerful actors into the policy field.

Inversely, a careful analysis of frame differences in a coalition of actors who seem to understand each other can expose coalition fault lines, or even unintended deviations in policy programmes. For example, Pijnappel has used frame analysis to show how a policy programme intended to develop alternatives for animal testing drifted away from the original understanding. The Dutch parliament wanted to reduce the number of animals used in experiments. As its policy was delegated to research foundations and implemented in research projects, the policy's mission was reframed as a problem requiring more research, specifically on physiological mechanisms underlying traditional animal tests. The researchers argued that, with a deeper understanding, they would be able to replace the old tests with more accurate ones. Ironically, this required *more* test animals, such as for validation of new tests against the old ones, and did not alter professional and legal standards that require animal tests. By reinterpreting the policy as a research problem requiring deeper understanding, the programme drifted off course (Pijnappel, 2016). Thus, frame analysis can help to put policies back on track and clarify problems of interpretation during policy implementation.

An implicit assumption in frame analysis is that an awareness of frames will lead to the possibility of reasonable deliberation, and hence to potentially improved agreement or change. This is not necessarily the case. Actors may have dug in, taking positions from which they cannot move without losing face. (This is one way to interpret the GMO controversy, where neither party can move without being perceived as losing the debate.) Alternatively, frames may have become so deeply embedded in institutions, with so much invested work, that change requires much more than just awareness (Star, 1991). Or powerful actors may maintain a dominant frame, even in the face of reasonable counter-arguments (Flyvbjerg, 1998). On a profound level, environmental policy making is never about reasonable deliberation alone, but always also about power, and hence political mobilisation will most likely be needed in addition to an awareness of assumptions or framing (Hoppe, 2010). While a reflection on frames may be an ingredient of transformation that leads to more durable and just environmental knowledge and policies, this is not likely to be enough in itself.

References

Brink, M. van den, (2010). *Rijkswaterstaan on the Horns of a Dilemma*. Delft: Eburon.

Douglas, M. (1986). *How Institutions Think*. Syracuse: Syracuse University Press.

Drenthen, M. (2017). Environmental Hermeneutics and the Meaning of Nature. In A. Thompson and S. M. Gardiner, eds., *Environmental Hermeneutics and the Meaning of Nature* (pp. 162–173). Oxford: Oxford University Press.

Flyvbjerg, B. (1998). *Rationality and Power: Democracy and Practice*. Chicago: University of Chicago Press.

Goffman, E. (1959). *The Presentation of Self in Everyday Life*. Garden City: Doubleday.

Hajer, M. (1995). *The Politics of Environmental Discourse: Ecological Modernization and the Policy Process*. Oxford: Clarendon Press.

Hajer, M., and Wagenaar, H. (2003). *Deliberative Policy Analysis: Understanding Governance in the Network Society*. Cambridge: Cambridge University Press.

Halffman, W., and Hoppe, R. (2005). Science/Policy Boundaries: A Changing Division of Labour in Dutch Expert Policy Advice. In S. Maasen and P. Weingart, eds., *Democratization of Expertise? Exploring Novel Forms of Scientific Advice in Political Decision-making* (pp. 135–152). Dordrecht: Kluwer.

Halffman, W., and Ragas, A. M. (2016). *Achter de Horizon: Omgaan met onzekerheid bij nieuwe risico's*. www.rijksoverheid.nl/documenten/rapporten/2017/08/23/beleid sperspectieven-voor-omgaan-met-onzekerheden-bij-nieuwe-risico-s

Halffman, W., and Ragas, A. M. (2018, forthcoming). Beyond the Horizon: Uncertainty in Environmental Policy for New Risks.

Hisschemöller, M., and Hoppe, R. (1995–1996). Coping with Intractable Controversies: The Case for Problem Structuring in Policy Design and Analysis. *Knowledge, Technology, and Policy*, 8(4), 40–60.

Hoppe, R. A. (2010). *The Governance of Problems: Puzzling, Powering, and Participation*. Portland: The Policy Press.

Jardine, N., Secord, J. A., and Spary, E. C., eds. (1996). *Cultures of Natural History*. Cambridge: Cambridge University Press.

Kreuter, M. W., Rosa, C. D., Howze, E. H., and Baldwin, G. T. (2004). Understanding Wicked Problems: A Key to Advancing Environmental Health Promotion. *Health Education & Behavior*, 31(4), 441–454.

Kwa, C. (1994). Modelling Technologies of Control. *Science as Culture*, 4(3), 363–391.

Lieshout, M. van, Dewulf, A., Aarts, N., and Termeer, C. (2011). Do Scale Frames Matter? Scale Frame Mismatches in the Decision Making Process of a 'Mega Farm' in a Small Dutch Village. *Ecology and Society*, 16(1), art. 38. www .ecologyandsociety.org/vol16/iss1/art38/

Metze, T. (2014). Fracking the Debate: Frame Shifts and Boundary Work in Dutch Decision Making on Shale Gas. *Journal of Environmental Policy & Planning*, 1–18. doi:10.1080/1523908X.2014.941462

Pijnappel, M. (2016). *Lost in Technification: Uncovering the Latent Clash of Societal Values in Dutch Public Policy Discourse of Animal-testing Alternatives*. PhD thesis. Nijmegen: Radboud University.

Rico, A., Van den Brink, P. J., Leitner, P., Graf, W., and Focks, A. (2016). Relative Influence of Chemical and Non-chemical Stressors on Invertebrate Communities:

A Case Study in the Danube River. *Science of the Total Environment*, 571, 1370–1382. doi:10.1016/j.scitotenv.2016.07.087

Scholten, P. (2008). *Constructing Immigrant Policies. Research-policy Relations and Immigrant Integration in the Netherlands, 1970–2004*. PhD thesis. Enschede: Twente University.

Schön, D., and Rein, M. (1994). *Frame Reflection: Toward the Resolution of Intractible Policy Controversies*. New York: Basic Books.

Scott, J. C. (1998). *Seeing like a State: How Certain Schemes to Improve the Human Condition Have Failed*. New Haven: Yale University Press.

Star, S. L. (1991). Power, Technologies and the Phonomenology of Conventions: On Being Allergic to Onions. In J. Law, ed., *A Sociology of Monsters: Essays on Power, Technology and Domination* (pp. 26–56). London: Routledge.

Stone, D. (1997). *Policy Paradox: The Art of Political Decision Making*. New York: Norton.

Verbrugge, L. N. H., Leuven, R. S. E. W., and Zwart, H. A. E. (2016). Metaphors in Invasion Biology: Implications for Risk Assessment and Management of Non-Native Species. *Ethics, Policy & Environment*, 19(3), 273–284. http://dx.doi.org/10.1080/21550085.2016.1226234

Yanow, D. (1996). *How Does a Policy Mean? Interpreting Policy and Organizational Actions*. Washington: Georgetown University Press.

Young, R. M. (1985). *Darwin's Metaphor: Nature's Place in Victorian Culture*. Cambridge: Cambridge University Press.

Case A Framing Climate Change

MIKE HULME

The case study on framing in the climate change debate demonstrates how profound variations in the understanding of climate change underlie some of the current disagreements. Various framing processes are at work in this example, including scale frames and metaphors. Mike Hulme shows how analysing these frames can clarify assumptions, as well as help to map what disagreement is about, such as by pointing to very different root causes of climate change. He shows how frames operate even in climate science and the world of climate models.

Many of the principles of framing can be illustrated using the concept of climate change. Scientific knowledge of climate and its changes has evolved rapidly over the last half century, and over the last three decades this has given rise to the public discourse and policy focus of climate change. Particularly influential in this regard has been the UN's Intergovernmental Panel on Climate Change (IPCC), a scientific assessment process which has exerted powerful framing influences on both scientific knowledge and policy responses. This case study explores how framing has been an inescapable part of climate science and the IPCC process, and of the discourse of climate change more widely, and points out some of the consequences of these framings.

Framing Climate Change

There are many ways to frame the phenomenon of climate change. Some frames engage certain audiences more than others, and some are more sugges-tive of certain types of policy interventions than others. As Chapter 3 has demonstrated, no frames – not even those that remain faithful to scientific facts – can be entirely neutral with respect to the effects that they generate on their audiences.

Take this framing:

A: The overwhelming scientific evidence tells us that human greenhouse gas emissions are resulting in global climate change that cannot be explained by natural causes. Climate change is real, humans are causing it, and it is happening right now.[1]

Fact. Nothing to challenge there. But how about this alternative frame:

B: The overwhelming scientific evidence tells us that human greenhouse gas emissions, land-use changes, and aerosol pollution are all contributing to regional and global changes in climate, which exacerbate the variability in climates brought about by natural causes. Because humans are contributing to climate change, it is happening now and in the future for much more complex reasons than previously in human history.

Frame B is also scientifically accurate, yet these two different framings of climate change open up very different possibilities for public and policy engagement with the issue. How climate change is framed shapes the possible responses. Frame B, for example, emphasises that human influences on climate are not just due to greenhouse gas emissions (and hence that climate change is not just about fossil energy use), but also result from land-use changes (greenhouse gas emissions and albedo effects) and from atmospheric aerosols (dust, sulphates, and soot). It also emphasises that these human effects on climate are as much regional as they are global. And it highlights that the interplay between human and natural effects on climate are complex, and that this complex entanglement of factors is novel.

Frame A, whilst scientifically accurate, is more partial and also more provocative. This framing might have the effect of reinforcing the polarisation of public opinion that exists around climate science, or it might be offered as a rhetorical move to generate a sense of urgency in policy negotiations. Yet there are important aspects of scientific knowledge about the climate system that accommodate more nuanced interpretations of uncertainty than are offered by Frame A and which open up more diverse sets of policy strategies. It is these aspects that Frame B seeks to foreground.

The IPCC's Framing of Climate

Since the IPCC was created in 1988, the knowledge it has presented to the world has been heavily framed using the paradigm of global climate modelling and

---

[1] This framing was taken from *The Conversation* blog post about a 2011 open letter on climate change, signed by Australia's 'top minds on the science behind climate change and on the efforts of "sceptics" to cloud the debate' (https://theconversation.com/climate-change-is-real-an-open-letter-from-the-scientific-community-1808).

Earth System Science (NASA, 1986).[2] Through five knowledge assessment cycles, the IPCC has conceptualised climate first and foremost as 'global', indexed iconically using the global-mean surface air temperature. This is a classic example of a 'scale-frame' (Van Lieshout et al., 2011; cf. Chapter 3, this volume), stressing the global rather than regional scale of the problem. Although the IPCC has never *formally* debated or adopted a unifying conceptual framework, its default position has always been to frame climate as a single, interconnected, physical system. The taken-for-grantedness of this powerful scale-frame was revealed by Thomas Stocker, the outgoing Co-Chair of Working Group 1 of the IPCC's *Fifth Assessment Report* (AR5). Commenting in 2015 on the possible structure of the future AR6, Stocker said: 'Many other opinions and suggestions have been aired. Regionalization of IPCC assessments is sometimes called for to give policy-makers and practitioners more and better regional information. In our view, this approach would undermine *the global character of the climate change problem exemplified by the IPCC*' (Stocker and Plattner, 2014, 165, emphasis added). This view of climate change as meaningful only in terms of its global character is one that the IPCC inherited in the late 1980s from the emerging Earth System Science community. Yet at the same time it is one that the IPCC, through its influence on the worlds of climate science and policy, has helped to shape. Once established, frames can be difficult to shift (see Chapter 3). Since the 1950s, the idea of 'climate' in Western science evolved from being predominantly interpretative, and hence geographically differentiated, to becoming enumerated, and hence readily globalised (Miller, 2004; Hulme, 2017). The era of satellites and computer models and a globally connected network of scientific institutions and standardised practices enabled this new framing of 'climate as global' to prevail. Rather than being framed as *regional, multiple*, and *situated in places* (e.g. how Alexander von Humboldt viewed climate in the nineteenth century; see Heymann, 2010), climate became framed as *global, singular*, and *placeless*. Regional climatic variations became interpreted through the framing narrative of *global* climate change, while global climate became the entity to be predicted by the new Earth System Science.

The effects of this global framing of climate are multiple. The flow of knowledge is established as being one-way (as too is causation; see Nielsen and Sejersen, 2012): *from* the global *to* the regional and, occasionally, *to* the local. Similarly, Working Group 3 of the IPCC – dealing with the mitigation of climate change – inherited the consequences of the IPCC's framing of climate as global. This working group evaluates global-scale studies that analyse technological, economic, and land-use intervention options to

---

[2] This section is based on parts of Turnhout et al. (2016).

achieve *global* objectives – for example, managing global carbon budgets or global temperature. Indeed, one of the clearest effects of the globalised framing of climate is the reification of global temperature. This indexed quantity – whether constructed from thermometer measurements, calculated from satellite retrievals, reconstructed from proxies, or modelled through computer code – has become central to the framing of climate change. The UN Framework Convention on Climate Change (UNFCCC) and the Paris Agreement of 2015 adopted this framing in their definition of dangerous climate change: warming that is no more than 2 degrees Celsius above the pre-industrial level, ideally no more than 1.5 degrees. Avoiding climatic danger, for humans and non-humans alike, is reduced to securing control of this global climatic index. Framing 'dangerous climate change' this way promotes solutions that are also global in scope, such as global carbon pricing or planetary geo-engineering, and deflects attention from other interventions that bring more direct regional benefits, such as improving air quality, enhancing human development, removing fossil fuel subsidies, and redesigning cities.

Yet there are critiques of this IPCC framing of climate as global. It has been argued that the global knowledge-making mode of the IPCC pays little attention to the multiple ways in which humans come to know their climates, or of how they live in places and imagine their futures (Jasanoff, 2010; Hulme 2017). These ways of knowing are embedded in local cultural practices and memories. The IPCC's appropriation of global knowledge means that place-based or indigenous knowledges become marginalised from the dominant centres and methodologies of global knowledge production and mobilisation (Ford et al., 2012). Another famous critique of this global framing of climate came back in 1991, in a pamphlet authored by Anil Agarwal and Sunita Narain. Their argument was that by framing climate as global, and changes in climate as caused by aggregated global emissions of greenhouse gases, differential responsibilities for these emissions were erased from the narrative. As Agarwal and Narain argued, 'luxury emissions' from the West were not on a par with 'survival emissions' from India (Agarwal and Narain, 1991). And it is true that, over the years, the framing of the climate change knowledge assessed by the IPCC has gradually evolved, such that in the Special Report on the 1.5-degree temperature target, due in 2018, one of the five framing chapters explicitly recognises climate change in the context of 'sustainable development, poverty eradication and reducing inequalities'.

Framing the Causes of Climate Change

The above discussion illustrates how climate is framed in the institutions and practices of the United Nations (the IPCC and UNFCCC). But framing the *causes* of climate change is even more contested than how the physical phenomenon of a changing climate per se is framed. How one frames a complex issue which connects social and material worlds – and all agree that, at the very least, climate change is complex – inevitably emphasises some aspects of that issue, whilst de-emphasising others. And, as with all frames, these emphases are not neutral. They result from judgements – whether careful or careless – made by those framing the issue, or they result from negotiations between different actors. The resulting frames have significant consequences for how public and policy audiences receive, interpret, and engage with the communication.

Take the case of the IPCC and the UNFCCC. Even these two related institutions of the United Nations frame the causes of climate change differently. For the IPCC, climate change 'may result from natural [. . .] processes or [. . .] forcings, or [from] persistent anthropogenic changes in the composition of the atmosphere or in land use'. But for the UNFCCC, climate change 'is attributed directly or indirectly to human activity that alters the composition of the global atmosphere and which is in addition to natural climate variability' (IPCC, 2001, 711). These two frames of climatic causation have significant political consequences, the UNFCCC focusing only on changes in climate that are additional to natural variability.

Beyond the scientific or diplomatic language of the IPCC and UNFCCC, we see a much more diverse set of causal frames at work in the wider public discourse. These frames are in effect narratives (see Chapter 3), meaningful stories which identify the key (human and non-human) protagonists of the issues at stake and their relationships with each other. To illustrate, consider the following five powerful frames[3] for the causes of climate change:

- market failure
- 'manufactured risk'
- global injustice
- overconsumption
- moral–social disorder

Framing climate change as caused by *market failure* is offered, for example, in the Stern Review of 2006: 'Climate change is the greatest example of market

---

[3] There are of course many more ways of framing the causes of climate change than these five. I merely offer these as salient examples of frames in public discourse.

failure the world has ever seen' (Stern Review, 2007, 1). It draws attention to a preferred set of policy interventions: those which seek to 'correct' the market by introducing pricing mechanisms for greenhouse gas emissions. This framing has been influential in establishing various types of market-based instruments in recent years, such as emissions trading and carbon pricing.

When climate change is framed as a *manufactured risk*, the focus is on the inadvertent downsides of ubiquitous fossil-energy-based technologies. It is suggestive of a policy agenda that promotes low- or zero-carbon technology innovation as the solution to climate change, as, for example, in the Breakthrough Energy Challenge (www.b-t.energy) that was launched just ahead of the 2015 Paris climate negotiations.

Radically different, however, is the frame of *global injustice*. Here, climate change is framed as the result of historical and structural inequalities in access to wealth and power, and hence unequal life chances. Climate change is caused by the rich and privileged exploiting the poor and disadvantaged, intimately linked in this frame to colonialism and capitalism (Klein, 2014). Any solutions to climate change that fail to tackle that underlying 'fact' are doomed to fail.

A related frame, but one with a different emphasis, is climate change as the result of *overconsumption*: too many (rich) people consuming too many (material) things. For this frame, policy interventions need to be much more radical than simply putting a price on carbon, promoting new clean energy technologies, or weakening the shackles of capitalism. The focus should either be on dematerialising economies (e.g. Simms, 2012) or on promoting human fertility management (e.g. Emmott, 2013).

Finally, a fifth frame would offer climate change as the result of a breakdown of *the moral–social order*. Thus, for example, Rudiak-Gould (2012) explains how Marshall Islanders understand climate change (*oktak in mejatato*) to be a result of 'improper relations' and a weakening of the normative social fabric. Common amongst non-Westernised cultures, framing climate change this way points to interventions which re-establish cultural–social norms that, at an ontological level, respect and stabilise natural processes.

These five frames around climate change all attract powerful audiences, interests, and actors in their support. All of these frames – with the partial exception of the fifth one – would be broadly consistent with the scientific knowledge assessed by the IPCC. And yet, because they are rooted in different ideologies and different understandings of the relationship between humans, technologies, and nature, they filter, interpret, and weigh that scientific evidence in different ways. The result is different ways of framing climate change that seem to promote certain forms of policy interventions over others.

The human influences on climate change – and the policy significance of these influences – are too complex to reduce public debate around climate change to binary caricatures: mainstream scientists versus climate sceptics, believers versus deniers, liberal progressives versus conservatives, activists versus quietists. There are multiple framings of the causes of climate change in which scientific evidence, attitudes to risk, political ideology, myths of nature, and so on, are deeply interwoven and entangled. Facts and values combine, through declaration or negotiation, to produce radically different framings (see Chapter 3, this volume). Being aware of these different framings, and reflecting on one's own preferences, beliefs, and ideologies, is essential for debating publicly and constructively how to respond to climate change.

Science Is Continually Reframing

Although the IPCC's framing of climate as global, singular, and placeless has been powerful and influential, we should also recognise that science, as a socialised form of inquiry into the world, is dynamic. Its own frames of investigation do not remain static. Prompted by events in the physical world, by changing political or funding regimes, or by changes in the social dynamics of scientific practice influenced by wider cultural change, there have been more or less subtle changes in how climate science has framed itself during this past quarter-century.

By this I do not mean the simple notion that climate or Earth System Science has been 'filling the gaps in knowledge' or 'narrowing the uncertainties' of climate predictions, which is how the IPCC in its first and second assessment reports in the 1990s described its future tasks. This would suggest that science progresses in a linear manner, always sure of its eventual object of under-standing, and only being thwarted in realising its goal of gaining a faithful account of reality 'as it really is' by incomplete theory, lack of observations, or inadequate simulation tools. Rather, we have observed how the philosophy and practice of climate science keeps changing the form and locus of the 'climate reality' being studied, through reframing its key questions, its forms of repre-sentation, its use of metaphor, and its public communications.

There are many examples of these changing frames at work. Since the founding of the IPCC in the late 1980s, changing scientific conceptions of the climate system have given much greater prominence to the role of non-linear thresholds and feedbacks. Whereas most of the graphs of future climate change in the first assessment report of the IPCC in 1990 were linear, the recognition and simulation of complexity in Earth System models has increasingly allowed the climate system to be framed in terms of tipping points, hysteresis loops, and dynamical thresholds. Indeed, the idea of 'tipping points' shows the power of a framing metaphor (see Chapter 3) to transform human conceptions of physical reality. Although used in the social and medical sciences for several decades, the 'tipping point' metaphor was only introduced into climate science in 2005, but has since become a dominant framing device in both science and policy advocacy (Russill, 2015) – and not always with predictable outcomes.

Another example of how changes in scientific philosophy and practice have altered the imaginative and practical implications of climate change is in descriptions of the future. Whereas in the 1980s and 1990s future climates were described in terms of scenarios (possibilities with unknown likelihood), more recent quantitative descriptions of future climate change have attached likelihoods or probabilities to the outcomes. These different forms of representation – different ways of framing the future – lead to very different psychologies of risk and to different philosophies of decision making. Indeed, the wider argument is that changes in how climate-modelling uncertainties are framed redefine how the idea of climate change is deemed relevant for adaptation and development strategies and for practical decision making. Recent years have therefore seen a shift away from a predict-and-adapt paradigm (which rests on ever more accurate and precise prediction) to one of robust decision making and decision support (which recognises the ineluctable uncertainty of the climatic future) (Weaver et al., 2013).

This paradigm change has itself been part of a wider reframing of the relationship between climate science and society. The dominant focus of climate science to develop and improve climate predictions on the timescale of decades-to-centuries is giving way to a new focus on seasonal-to-decadal prediction (Keenleyside and Ba, 2010). This change in how modelling strategies are framed has implications for policy discourse by changing the relative weights given to different causal explanations of climatic change and climate variability. Whereas long-lived greenhouse gases dominate centennial-scale climate, a subtle mix of natural factors and short-lived radiative forcing agents are more important on seasonal-to-decadal timescales and on shorter-term 'weather'. This reframing of climate science carries implications for institutional design and policy formation. The idea of 'climate services' is gaining ground through initiatives such as the World Meteorological Organization's Global Framework for Climate Services (Vaughan and Dessai, 2014). Rather than being driven by questions concerning detection and attribution of human influences and long-term prediction of climate, climate science is being directed to provide climate information that can 'facilitate climate-smart decisions that reduce the impact of climate-related hazards and increase the benefit from benign climate conditions' (Hewitt et al., 2012, 831).

And the whole field of climate attribution studies is also changing the priorities of climate science. Rather than asking the question which dominated the 1990s and 2000s – 'have we detected human influence on the climate system?' – climate scientists are now investigating a rather different question: 'what is the probability that this extreme weather event was caused by human greenhouse gas emissions?' Here, too, changes in scientific framing have the potential to reconfigure political interests around climate change. The investigation of weather event attribution opens up the possibility of a new policy discourse around international legal liability for weather loss and damage, although there remain many unresolved difficulties with this manoeuvre (Hulme et al., 2011).

Conclusion

There are many ways of framing the political, economic, and cultural challenges of a changing climate, and the above examples show how the changing paradigms and emphases of scientific research carry influence in this struggle for framing. The 'handles' offered by scientific knowledge for understanding the phenomenon of climate change keep changing. Or, to put it another way: what climate change *means* for humans cannot simply be read from a scientific script; it has to be constructed, normatively, using wider sets of resources, imaginaries, and commitments. Facts and values become mutually embedded. This is a contested process. And in this struggle for meaning, how scientific knowledge is 'framed' matters profoundly.

References

Agarwal, A., and Narain, S. (1991). *Global Warming in an Unequal World*. Delhi: Centre for Science and the Environment.

Emmott, S. (2013). *Ten Billion*. New York: Vintage.

Ford, J. D., Vanderbilt, W., and Berrang-Ford, L. (2012). Authorship in IPCC AR5 and its Implications for Content: Climate Change and Indigenous Populations in WGII. *Climatic Change*, 113(2), 201–213.

Hewitt, C., Mason, S., and Walland, D. (2012). The Global Framework for Climate Services. *Nature Climate Change*, 2(12), 831–832.

Heymann, M. (2010). The Evolution of Climate Ideas and Knowledge. *WIREs Climate Change*, 1(4), 581–597.

Hulme, M. (2017). *Weathered: Cultures of Climate.* London: SAGE Publications.

Hulme, M., O'Neill, S. J., and Dessai, S. (2011). Is Weather Event Attribution Necessary for Adaptation Funding? *Science* 334, 764–765.

IPCC, Intergovernmental Panel on Climate Change (2001). *Climate Change 2001: Mitigation. A Report of Working Group III of the Intergovernmental Panel on Climate Change.* www.ipcc.ch/ipccreports/tar/wg3/pdf/WGIII_TAR_full_report.pdf

Jasanoff, S. (2010). A New Climate for Society. *Theory, Culture & Society,* 27(2/3), 233–253.

Keenleyside, N. S., and Ba, J. (2010). Prospects for Decadal Climate Prediction. *WIREs Climate Change,* 1(5), 627–635.

Klein, N. (2014). *This Changes Everything: Capitalism vs the Climate.* New York: Simon & Schuster.

Lieshout, M. van, Dewulf, A., Aarts, N., and Termeer, C. (2011). Do Scale Frames Matter? Scale Frame Mismatches in the Decision Making Process of a 'Mega Farm' in a Small Dutch Village. *Ecology and Society,* 16(1), art. 38: www.ecologyandsociety.org/vol16/iss1/art38/

Miller, C. A. (2004). Climate Science and the Making of a Global Political Order. In S. Jasanoff, ed., *States of Knowledge: The Co-production of Science and Global Order* (46–67). London: Routledge.

NASA (1986). *Earth System Science: A Program For Global Change.* NASA Earth System Science Committee.

Nielsen, J. O., and Reenberg, A. (2012). Exploring Causal Relations: The Societal Effects of Climate Change. *Danish Journal of Geography,* 112(2), 89–92.

Rudiak-Gould, P. (2012). Promiscuous Corroboration and Climate Change Translation: A Case Study from the Marshall Islands. *Global Environmental Change,* 22(1), 46–54.

Russill, C. (2015). Climate Change Tipping Points: Origins, Precursors, and Debates. *WIREs Climate Change,* 6(4), 427–34.

Simms, A. (2012). *Cancel the Apocalypse: The New Path to Prosperity.* London: Little, Brown.

Stern Review (2007). *The Economics of Climate Change: The Stern Review.* Cambridge: Cambridge University Press.

Stocker, T. F., and Plattner, G. K (2014). Rethink IPCC Reports. *Nature,* 513, 163–165.

Turnhout, E., Dewulf, A., and Hulme, M. (2016). What Does Policy-Relevant Global Environmental Knowledge Do? The Cases of Climate and Biodiversity. *Current Opinions in Environmental Sustainability,* 18, 65–72.

Vaughan, C., and Dessai, S. (2014). Climate Services for Society: Origins, Institutional Arrangements, and Design Elements for an Evaluation Framework. *WIREs Climate Change,* 5(5), 587–603.

Weaver, C. P., Lempert, R. J., Brown, C., Hall, J. A., Revell, D., and Sarewitz, D. (2013). Improving the Contribution of Climate Model Information to Decision-Making: The Value and Demands of Robust Decision Frameworks. *WIREs Climate Change,* 4(1), 39–60.

4

Science, Politics, and the Public in Knowledge Controversies

ESTHER TURNHOUT AND THOMAS GIERYN

4.1 Introduction

Controversies are an inevitable part of current science–policy–society relations. Getting people to agree on the facts of nature and environment has never been a smooth process. While some of these controversies can be seen as scientific or technical discussions – for example, about methods or data and their interpretation – some others have assumed a very public character. Well-known examples outside the environmental domain are the controversies over evolution versus creationism, vaccination, and the health effects of living near radio towers. Within the environmental domain there are persistent knowledge controversies over genetic modification (GM), climate change, and numerous other issues in natural resource management. In this chapter, we discuss what forms of reasoning underlie controversies; the extent to which controversies are never only about facts or knowledge but also, simultaneously, about policy and society; and how they are settled in practice. We use the example of controversies to illustrate some general patterns and draw lessons that are more widely relevant for understanding science–policy–society interactions.

The history of science involves a number of examples that illustrate how and why knowledge controversies emerge, and also how they are eventually settled. Many of these stories are about the question of how credible facts can be produced. For example, Galileo's discovery that the moon had an uneven surface was very much the result of his use of the newly invented telescope. Similarly, one of the arguments against the claimed existence of this uneven surface was why we should trust the observations made with this new instrument over those made with our own eyes. Although in modern times this is difficult to imagine, we have to realise that this is not an irrational argument. Think about the Higgs Boson particle, for example, whose discovery cannot be imagined without the existence of the highly elaborate machinery of the

European Organization for Nuclear Research (CERN). Knowledge controversies also concern the question of *where* to look for the truth. When genetic analysis started to emerge as a method to classify plants according to phylogenetic relationships rather than morphological characteristics, as was hitherto customary according to the Linnaean taxonomy, a debate ensued about what exactly constitutes a plant species and where species' essential characteristics were located (Dean, 1979).

Many controversies, including the ones mentioned above, eventually get settled. However, when they do it is often not – or at least not only – because incontrovertible proof has emerged. Oreskes (2004) discusses the case of plate tectonics, a theory that when it was first introduced proved very difficult to accept. Opponents pointed to the absence of empirical proof – which at that time was impossible to obtain because it was not possible to measure the movement of the plates – as the main reason to reject the theory. In the end however, the absence of proof did not prove to be an insurmountable obstacle, and the theory was accepted long before empirical proof could finally be obtained. These examples of knowledge controversies illustrate three main features of knowledge controversies that are currently also recognisable in the environmental domain. The first observation is that these controversies often focus on the processes and methods by which contested facts are produced. The second is that there is a social context to these controversies that influences which facts are readily accepted and which are contested. The third observation is that knowledge controversies are often settled not by means of evidence, but because people choose one side or the other, or disengage from the issue.

Looking at science from the perspective of controversies has proven to be very helpful for understanding science. When science is contested, the notion and definition of science become problematised and the relation between science and society is openly debated. Thus, controversies reveal things about science that normally remain hidden. Therefore, it is no surprise that the history and sociology of science include many cases that show that science has never been free of conflict and contestation. This is not necessarily a problem for science or for society; as was explained in Chapter 2, some would even argue that critical scrutiny of claims is a core value of science. There is, however, a large variety in kinds of controversies. On a relatively simple level, you can say that there are 'small' controversies, which can rather be seen as scientific disagreements and disputes, and which eventually get resolved. However, some controversies do not get resolved and flow over into society. These persistent knowledge controversies typically involve a wide variety of actors: policy makers and politicians, interest representatives from industry, citizen groups, and environmental organisations are often involved. In addition, the media – including social media – increasingly play an important, and often catalysing, role in knowledge controversies. It is those kinds of knowledge controversies that we focus on in this chapter.

It is important to note that these 'bigger' controversies are not just about knowledge; contestations over knowledge are entwined with contestations over the potential political and societal implications of that knowledge. Decisions about what measures to take to combat climate change, whether to teach creationism and evolution in schools, or whether to permit genetically modified organisms (GMOs) in food are directly related to scientific knowledge about these issues. Thus, these controversies are not just about facts, but also simultaneously about values and interests, and this may very well be an important reason why they have proven so difficult to solve. As we will discuss in detail in this chapter, the typical response for settling controversies by generating more research until certainty about facts can be obtained which will result in consensus about courses of action in environmental policy – called the linear model – is misguided in these cases. While this way of thinking seems sensible and can work well in some cases, it is counterproductive in the kind of knowledge controversies discussed in this chapter. By ignoring the entwinement of facts, values, and interests, the linear model fuels rather than resolves knowledge controversies. Before arriving at a clear understanding of productive ways forward at the end of this chapter, we will first explain in detail the assumptions and shortcomings of the linear model.

4.2 Speaking Truth to Power and the Linear Model of Science–Society Relationships

As we have outlined in the preceding chapters, the relation between science, society, and decision making is often complex. To an important extent, difficulties in these relationships arise from deeply embedded ideologies and beliefs about the nature of science, such as those discussed in Chapter 1, and about the way science and society should relate. Specifically, scientific processes of knowledge production are associated with values of neutrality, objectivity, and truth. Societal processes of decision making, on the other hand, are often associated with subjectivity, power, and taking sides. According to these standard images, science and decision making are defined as each other's opposites (see Table 4.1). From this perspective, it is not surprising that when science and society meet, problems occur. In light of such incompatible

Table 4.1: *Standard images of science and decision making (adapted from Huitema and Turnhout, 2009)*

Decision making	Science
Values	Facts
Interests	Independent knowledge
Subjective	Objective, neutral
Ideology and power	Truth
Bargaining and negotiation	Standardised methodology
Support and acceptance	Reliability and validity
Action and achievement	Explanation and understanding
Cunning	Reason

perspectives and conflicting interests, scientists and decision makers will inevitably fail to understand each other and will be unable to communicate or cooperate effectively.

To scientists, close connections to society and decision-making processes may be seen as corrupting core scientific values. Scientists typically complain about having to simplify their analysis, change inconvenient findings, or about a lack of attention for important new insights. Inversely, decision makers complain about scientists' failure to appreciate societal concerns or the political need to balance various interests or raise support for decisions. At the same time, though, there is a clear desire to use scientific knowledge in decision making for environmental issues. The dominant way of thinking about the appropriate relationship between science and decision making has been characterised as 'speaking truth to power' (Wildavsky, 1979; Ezrahi, 1990) and can be visualised as in Figure 4.1.

Although at first sight this linear model may seem attractive, it is in fact a simplistic and naïve understanding of both scientific and decision-making processes. Firstly, decision making is driven by a variety of legitimate factors, including, for example, the balancing of societal interests, and scientific knowledge is often only one of the sources to inform decision making. Second, scientific knowledge does not simply translate into action; this requires interpretation, a way of coping with inevitable uncertainties, and making sense of competing knowledge claims. Thirdly, science itself does not operate separate from societal concerns; different kinds of societal factors – power, values, interests, beliefs – shape the production of scientific knowledge and, as such, science does not just speak to power, it also contains and is shaped by power. Despite these problems, the linear model functions in society as a powerful ideal that many actors in society and in science adhere to.

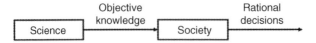

Figure 4.1 The linear model of science–society relationships.

Before we illustrate how this ideal operates in knowledge controversies, we first need to discuss it in a more elaborate way. What this model communicates is, first, that scientific knowledge is very important for rational decision making and, second, that scientific knowledge unproblematically translates into rational decisions: 'if we all agree on the facts we will know what to do'. Conversely, if decision makers are seen to make the wrong decisions or if they fail to take effective action, this must be due to a lack of knowledge or a lack of understanding of science. In other words, when actors feel that unfounded decisions are made, according to the linear model it follows that this must be due to a gap in the relation between science and policy. While policy has a societal function and policies have to be made and legitimised in the public domain, the linear model ignores this and instead promotes a technical framing of environmental issues where both the definition of the problem and the scope of possible solutions are placed in the expert domain. In doing so, it unduly puts science in a central position in environmental decision making. This has resulted in the simultaneous scientisation of policy and the politicisation of science (Weingart, 1999). Scientific knowledge is considered so important in underpinning decision making that we can say that environmental policy making has become completely scientised. At the same time, this creates strong incentives to politicise science.

We continue this chapter by zooming in on knowledge controversies in environmental politics and policies and on knowledge controversies that involve the public. These are not distinct types of controversies, but they allow us to discuss the different dynamics at play in knowledge controversies.

4.3 Science, Policy, and Politics in Knowledge Controversies

As explained earlier, much of current environmental policy has become scientised to such an extent that it has become crucially important to 'have the facts on your side'. Consequently, when there is disagreement about the preferred course of action in policy and decision making, it is not surprising to see debates starting to focus on the scientific facts that underpin these different preferences, the way in which these facts were produced, and by whom. In these situations, we can speak of the 'politicisation' of science:

science becomes a resource in political struggles. We can illustrate this with the example of cockle fisheries in the Wadden Sea (Turnhout et al., 2008). The Wadden Sea is an important nature conservation area. It enjoys a degree of protection, but human use is also allowed. One of these activities was cockle fishing. For many years, there have been debates about the ecological and environmental damage of this form of fishery. Specifically, there was one coalition that consisted of ecologists and nature conservation NGOs that argued that cockle fisheries are highly damaging and should be prohibited. The other coalition, consisting of fisheries scientists and fisheries organisations, argued that these claims were exaggerated and that cockle fisheries did not cause irreversible damage to the ecosystem. Both coalitions made scientific claims pertaining to the extent of damage caused by cockle fisheries. They also used scientific arguments to dismantle the knowledge claims of the opposing coalition, for example by pointing to uncertainties or allegedly incorrect methodologies. Thirdly, they attempted to discredit the scientific knowledge used by the opposing coalition by claiming that the other scientists were advocating their political positions in favour of conservation or fisheries, and that they were therefore not independent or objective. In the end, it was not scientific procedures but the court of justice that settled the controversy by deciding to close the fisheries. As the analysis of climate change and the IPCC (Case B: What Does 'Climategate' Tell Us) illustrates, climate science is another example in which different scientific and other actors accuse each other of not doing proper science, of being advocates of either the traditional oil companies or the environmental movement, and of failing to remain objective and independent.

What we see here is that stereotypical images of science (such as objectivity, disinterestedness, or truth; see Table 4.1) now appear not as universally valid demarcation criteria, but as rhetorical resources that are mobilised by actors in disputes to claim scientific status or reject unwelcome knowledge claims. Uncertainties also feature in these debates. Since scientific knowledge can never claim full certainty, it is in principle always possible to use these uncertainties to dismiss scientific knowledge (see Chapter 5). We should recognise that what results from this reliance on science to solve problems, and the concomitant controversies with their strategies of using science to legitimise and delegitimise different preferences and views, is often far removed from what can be considered as rational decisions. Thus, as Collingridge and Reeve (1986) have argued, knowledge that fits certain positions and preferences is generally not criticised and accepted, while knowledge that goes against these positions and preferences is critically scrutinised and rejected. This has led Sarewitz (2004) to conclude that while science is often believed capable of settling controversies, in practice it often makes them worse. While this is perhaps an overly cynical conclusion, it does point to the importance of context in knowledge controversies: in practice, depending on the context, different quality standards, criteria, and definitions of science are used to evaluate the validity of knowledge claims. Gilbert and Mulkay's (1984) study of a controversy in microbiology shows how scientists selectively use two different repertoires. The first repertoire, containing arguments about objectivity and facts, is used to argue for the validity and truth value of the claims they support. The second repertoire, which exposes the values and contingencies of knowledge production, is used to discredit the claims that they disagree with. In a similar vein, Nelkin (1982, 191) writes the following about the debate about creationism, or intelligent design, and evolution: 'Biologists and creationists alike claim the other bases its beliefs on faith; each group argues with passion for its own dispassionate objectivity; and each bemoans the moral, political and legal implications of the alternative ideology.'

These studies reveal a remarkable asymmetry in how knowledge controversies are often interpreted: only the side that is assumed to have got it wrong is criticised, while the side that is assumed to have got it right is believed to have used appropriate scientific methods and has hence discovered the truth. The history of science includes many examples of cases where only those forms of science that are currently predominantly conceived as wrong are claimed to be corrupted by politics or ideologies (famous historical examples include Lysenkoism (Joravsky, 1970) and phrenology (Shapin, 1979)). The 'merchants of doubt' narrative discussed in the case about climate change and the IPCC (Case B: What Does 'Climategate' Tell Us) offers a modern example of such an asymmetrical reading, whereby only one side is assumed to be politically motivated to protect the interests of the fossil fuel industry and discredit climate science. However, it would be wrong to assume that the science that we believe to be true is completely value free. While with hindsight, or reasoned from certain convictions or beliefs, it may be easy to say who is right and who is wrong, often this is not immediately evident, especially when both sides appear to make plausible claims about the credibility of their facts. It is important not to prematurely assume who is right and who is wrong, and to take seriously the social context in which a knowledge controversy plays out. This context affects all parties in the debate, not just the ones that will end up on the wrong side of history. In other words, we need to look at the different sides of the controversies with a symmetrical perspective.

4.4 Science and the Public in Knowledge Controversies

As explained earlier in this chapter, the linear model views science as an important requirement for rational decision making. And although the problems of the linear model are well established, it still figures as an important ideal in practice. This model assumes that in cases where public views about the desirability of science and technology do not match those of the experts, one assumption is that the public must suffer from an information deficit (Irwin and Wynne, 1996). Again reasoned from the linear model, the suggested solution is equally simple: improving societal practices and decisions then becomes a matter of improving the public understanding of science and hence of improving the communication between science and society.

Later chapters in this book (see, for example, Chapter 6) will delve further into the question of how we improve the connection between science and decision making, and how can we generate scientific knowledge that is legitimate and relevant for policy processes. For now, we need to consider more critically how the information deficit model conceives the public. A number of issues have emerged recently where science has been confronted with public resistance. Examples include the introduction of genetic modification in agriculture and food, and vaccination campaigns. Typical narratives about contestations between scientists and public groups will, in line with the assumptions of the information deficit model, assert that public resistance can be explained by referring to a lack of understanding on the side of the public or by referring to the emotions that supposedly drive public responses. A clear example of this is a quotation by Paul Nurse in a BBC television programme called 'Science under Attack' (Jay, 2011):

> The GM debate raises the issue of public trust in science. The research has to be carried out under strict security because feelings are still running high . . . Part of the problem is [that] the mainstream science has failed to win over the public. There is a gap between the fears of some sections of the public and the opinion of scientists that what they are doing is both useful and safe.

Here we see the public cast in a rather dismissive way as being driven by feelings and fear, while science is cast as mainstream (to underline its authority), useful, and safe. The quotation also suggests that the lack of trust in

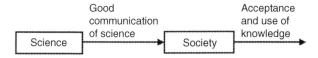

Figure 4.2 The information deficit model.

science by the public is caused by a lack of knowledge on the usefulness and safety of what scientists are doing. In line with the information deficit model, the only job for science is to win over the public by means of good communication of science. The case study of fracking (Case C: Whose Deficit Anyway?) offers an insightful analysis of the way in which public resistance is framed in terms of the information deficit model.

Many studies have shown that there are problems with this reasoning. First of all, in many cases, citizens or citizen groups do possess expert knowledge about the subject at hand, but this does not automatically mean that they agree with scientists about the facts, about the interpretation of facts, or about the consequences of these facts. Furthermore, gaining expert knowledge may result in stronger public resistance against certain forms of science and associated practices. Epstein's (1996, 2) account of AIDS research demonstrates how groups of AIDS activists attained expert knowledge and started to criticise biomedical research for not taking AIDS seriously and for not being 'conducive to medical progress and the health and welfare of their constituency'. Importantly, even if citizens can be shown to lack expert knowledge about the scientific details of genetic engineering, for example, this does not mean that their attitude towards this technology is uninformed or irrational. This attitude may be based on previous experiences wherein scientific experts and their institutions acted in a untrustworthy way. The case study of fracking (Case C: Whose Deficit Anyway?) and the example of sheep farming in the English Lake District (Chapter 8) offer good illustrations of this. Another possible explanation is that the public's beliefs may be based on different conceptions of food safety that do not fit with scientific framings of risk and safety. Wynne (2001) has used the example of GM to argue that framing environmental issues in terms of risk privileges specific science-based techniques to assess those risks in quantitative terms, alienates citizens who evaluate safety and danger in a different way, and dismisses their concerns as mere perceptions of risk which are driven by fear or emotion and which can be juxtaposed against the 'real' risks as science has calculated them.

Much like we have seen in the previous section, any course of action to enhance public understanding of science or trust in science that is based on the assumptions of the information deficit model is likely to backfire. Greater knowledge of the issue at hand may lead to greater resistance – for example, because it triggers a more acute appreciation of the dangers and uncertainties of certain technologies, or because it strengthens the view of citizens that scientific experts are unable to speak to their concerns. A crucial issue here is trust. Trust is a condition for accepting scientific knowledge claims as credible. However, studies in the public understanding of science have demonstrated that trust depends on much more than the actual truth value of claims; it also depends on the apparent trustworthiness of the behaviour of scientific experts and their institutions, and on the way in which science addresses and resonates with the concerns of citizens. Thus, more and better scientific knowledge, or better communication of science alone, will not restore public trust in and acceptance of science. Instead, what are required are more active forms of engagement of publics with science and of science with publics in order to build trust and address the ways in which issues are framed.

Building trust is not something that is readily accepted as a core task of science. Because the facts are assumed to speak for themselves, the general belief is that trust is not required. However, Chapters 2 and 3 have made clear that that assumption cannot be upheld. Even more so, objective information is often seen as a way to replace a dependence on other forms of authority that are often considered undesirable (e.g. hierarchy) (Porter, 1995). For example, systems of performance measurement and evaluation have emerged to check policy makers, companies, and an increasingly wide variety of other actors, because we do not want to rely on trust alone. And these systems of performance measurement, in turn, are assumed to enhance the trustworthiness of these actors. Another example is sustainability certification in food production, where information about the way in which food is produced is assumed to build trust. However, as Porter (1995) shows, the relationship between objective information, trust, and authority is complex. He has documented a number of historical cases that show that scientific disciplines that are under scrutiny (including engineering and economics) exhibit the strongest drive towards objectivity, while disciplines that traditionally enjoy high authority (e.g. high energy physics) rely much less on mechanisms and procedures for objectivity. This suggests that a strong emphasis on objectivity does not necessarily result in authority; rather, it could be seen as an indicator for a lack thereof. So, what we see is complex interdependencies of trust, authority, and objective scientific facts. If there is authority and trust, objectivity is often not required. However, if there is a lack of trust and authority, objective scientific facts will not be effective, because the acceptance of these facts itself depends on trust (Shapin, 1995).

4.5 Making Sense of Knowledge Controversies

In this chapter, we have discussed knowledge controversies and the relation between science, policy, and the public. We have seen that in cases of controversy where getting the facts on your side is an important strategy, questions of what science is, what it means to be a good scientist, and who should be counted as a good scientist are actively debated and disputed. We have also seen that the assumptions of the linear model and its twin, the information deficit model, are problematic for four reasons: 1) it makes knowledge too important as an input for decision making; 2) it ignores that the decisions of knowledge users – policy makers, citizens – are based on legitimate factors other than science alone; 3) it fails to recognise that scientific knowledge does not unequivocally translate into action; and 4) it ignores the extent to which science and society are entwined in the sense that science is influenced by societal and policy concerns and therefore does not just 'speak to power', but is shaped by and carries power itself. Consequently, solutions based on this model alone are unlikely to be effective in resolving knowledge controversies and may even fuel them. Instead, we have discussed three crucial factors in under-standing knowledge controversies:

1. *Context matters*. In different circumstances, different criteria and arguments are used to accept or reject scientific knowledge claims. In some cases evidence for a claim is important; in other cases a plausible theory suffices. In some cases close collaboration with policy makers is a reason to dismiss science; in other cases it enhances the credibility of knowledge.
2. *Framing matters*. How certain issues *get* framed as controversies, and how they *are* framed as controversies, matters for how problems and suggested solutions are seen. When a knowledge controversy is framed as being caused by political motivations, we see that actors will try to expose these political motivations and promote the use of neutral and objective science. If the controversy is framed as being caused by an uninformed and emotional public, we see a tendency to educate the public about the real risks as calculated by science.
3. *Trust matters*. Trust is a condition for accepting scientific claims as credible. Thus, if for some reason trust in science is lacking, more science or better communication of science will not help, and may even do harm.

4.6 Building Trust

In this concluding section, we will reflect on the implications of knowledge controversies and possible strategies to address them. First, it is clear that these controversies can be quite damaging to the reputations of the scientists involved. Alleging that (s)he is not neutral, or is an advocate for the environ-mental movement or for industry, is about the worst thing you can say to a scientist who cherishes values of independence, disinterestedness, and object-ivity. Although, arguably, some scientists enter into these controversies will-ingly and consciously, it this is not always something that can be avoided. When environmental issues are framed as technical, scientific knowledge, the scientists that produced this knowledge – willingly or unwillingly – become part of the debate (In't Veld, 2000; Moore, 2008).

It must be recognised that knowledge controversies are inevitable. Whatmore (2009) explains that they have a productive quality to them and can be seen as events, which may give rise to new framings, insights, ideas, and knowledge. In light of this, we can think of strategies that may be able to avoid the kinds of stalemate and polarised debates that we have discussed in this chapter. Using the insights presented in this chapter, these strategies have to recognise 1) the multidimensional nature of knowledge controversies that goes beyond a mere technical framing of what the issue at stake is; and 2) the different – legitimate – reasons that knowledge users may have to either use, trust, or reject different forms of knowledge. These two assertions imply a fundamental move away from many of the deeply embedded assumptions of the linear model, and strategies based on them will not be easy to put into practice. Nevertheless, they may help build the trust that is needed for science to play a legitimate role in environmental decision making.

One example of such an attempt has been Ravetz's initiative to organise a meeting consisting of climate scientists and so-called climate sceptics (Ravetz, 2011). The aim of this meeting was to have open deliberation about different viewpoints and knowledge claims and come to some sort of reconciliation. Although the workshop was attended by a high number of key actors and was set up to promote non-violent dialogue, the expected reconciliation did not occur. In the highly volatile context of the debate, right in the aftermath of 'climategate' (also see Case B: What Does 'Climategate' Tell Us), it proved too difficult for the different sides in the debate to come together or admit any common ground. In this context, apparently, searching for middle or common ground or attempting to recognise and understand each other's positions was not possible, and the only available option was to reproduce the existing polarised positions. Maarten Hajer's strategy, as the director of the Dutch Environmental Assessment Agency, of addressing the Dutch errors in one of the IPCC reports using principles and theories from deliberative democracy faced similar challenges (Hajer, 2012a; Hajer, 2012b; Tuinstra and Hajer, 2014). Mainstream scientists felt that by engaging climate sceptics, the agency risked its scientific reputation. Similarly, climate sceptics were not always keen to be involved because their credibility and authority also depended on main-taining distance. Thus, dominant ideals about science and about science–

society relations derived from the linear model continued to play a role and prevented the success of attempts to build trust.

However, there are also examples with different outcomes. One of these relates to flood risk management (Landström et al., 2011; Lane et al., 2011). These authors show that a process of deliberation and dialogue using a companion modelling approach (also see Chapter 7) was able to build trust and jointly create new flood knowledge that was able to reconcile the differing views of experts and other knowledge holders and make a positive contribution to local flood management practices. The case study of Loweswater (Case I: The Loweswater Care Project) offers a similar example.

Although it is clear that the reconciliation of knowledge controversies has to focus not just on getting the facts right but also, and at the same time, on building trust, these examples show that this is never easy, and is sometimes even impossible because the assumptions and ideals of the linear model are deeply rooted and breaking away from them is often risky.

References

Collingridge, D., and Reeve, C. (1986). *Science Speaks to Power: The Role of Experts in Policy Making*. London: Frances Pinter Publishers.

Dean, J. (1979). Controversy over Classification: A Case Study from the History of Botany. In B. Barnes and S. Shapin, eds., *Natural Order, Historical Studies of Scientific Culture* (pp. 211–230). London: SAGE Publications.

Epstein, S. (1996). *Impure Science: AIDS, Activism, and the Politics of Knowledge*. Oakland: University of California Press.

Ezrahi, Y. (1990). *The Descent of Icarus: Science and the Transformation of Contemporary Democracy*. Cambridge: Harvard University Press.

Gilbert, G. N., and Mulkay, M. (1984). *Opening Pandora's Box: A Sociological Analysis of Scientists' Discourse*. Cambridge: Cambridge University Press.

Hajer, M. (2012a). A Media Storm in the World Risk: Enacting Scientific Authority in the IPCC Controversy (2009–10). *Critical Policy Studies*, 6(4), 452–464.

Hajer, M. (2012b). Living the Winter of Discontent: Reflections of a Deliberative Practitioner. In M. Heinlein, C. Kropp, J. Neumer, A. Poferl and R. Römhild, eds., *Futures of Modernity: Challenges for Cosmopolitical Thought and Practice* (pp. 77–94). Bielefeld: Transcript.

Huitema, D., and Turnhout, E. (2009). Working at the Science–Policy Interface: A Discursive Analysis of Boundary Work at the Netherlands Environmental Assessment Agency. *Environmental Politics*, 18(4), 576–594.

In't Veld, R. J. (2000). *Willingly and Knowingly: The Roles of Knowledge about Nature and the Environment in Policy Processes*. The Hague: Advisory Council for Environmental Research (RMNO).

Irwin, A., and Wynne, B. (1996). *Misunderstanding Science? The Public Reconstruction of Science and Technology*. Cambridge: Cambridge University Press.

Jay, E. (2011, January 24). *Horizon 2010–2011: Science Under Attack*. BBC TV. www.bbc.co.uk/programmes/b00y4yql

Joravsky, D. (1970). *The Lysenko Affair*. Cambridge: Harvard University Press.

Landström, C., Whatmore, S. J., Lane, S. N., Odoni, N. A., Ward, N., and Bradley, S. (2011). Coproducing Flood Risk Knowledge: Redistributing Expertise in Critical 'Participatory Modelling'. *Environment and Planning A*, 43(7), 1617–1633.

Lane, S. N., Odoni, N., Landström, C., Whatmore, S. J., Ward, N., and Bradley, S. (2011). Doing Flood Risk Science Differently: An Experiment in Radical Scientific Method. *Transactions of the Institute of British Geographers, New Series*, 36, 15–36.

Moore, K. (2009). *Disrupting Science: Social Movements, American Scientists, and the Politics of the Military, 1945–1975*. Princeton: Princeton University Press.

Nelkin, D. (1982). *The Creation Controversy: Science or Scripture in the Schools*. New York: Norton

Oreskes, N. (2004). Science and Public Policy: What's Proof Got To Do with It? *Environmental Science & Policy*, 7(5), 369–393.

Porter, T. M. (1995). *Trust in Numbers: The Pursuit of Objectivity in Science and Public Life*. Princeton: Princeton University Press.

Ravetz J. R. (2011) 'Climategate' and the Maturing of Post-normal Science. *Futures*, 43, 149–157.

Sarewitz, D. (2004). How Science Makes Environmental Controversies Worse. *Environmental Science and Policy*, 7(5), 385–403.

Shapin, S. (1979). The Politics of Observation: Cerebral Anatomy and Social Interests in the Edinburgh Phrenology Disputes. *The Sociological Review*, 27(1), 139–178.

Shapin, S. (1995). Cordelia's Love, Credibility and the Social Studies of Science. *Perspectives on Science*, 3(3), 225–271.

Tuinstra, W., and Hajer, M. (2014). Deliberatie over klimaatkennis: de publieke omgang van het PBL met IPCC-fouten en klimaatsceptici. *Bestuurskunde*, 23(2), 38–45.

Turnhout, E., Hisschemöller, M., and Eijsackers, H. (2008). Science in Wadden Sea Policy: From Accommodation to Advocacy. *Environmental Science and Policy*, 11 (3), 227–239.

Weingart, P. (1999). Scientific Expertise and Political Accountability: Paradoxes of Science in Politics. *Science and Public Policy*, 26(3), 151–161.

Whatmore, S. J. (2009). Mapping Knowledge Controversies: Science, Democracy and the Redistribution of Expertise. *Progress in Human Geography*, 33(5), 587–598.

Wildavsky, A. (1979). *Speaking Truth to Power: The Art and Craft of Policy Analysis*. Boston: Little, Brown and Company.

Wynne, B. (2001). Creating Public Alienation: Expert Cultures of Risk and Ethics on GMOs. *Science as Culture*, 10(4), 445–481.

Case B What Does 'Climategate' Tell Us About Public Knowledge Controversies

SILKE BECK

Climate change is one of the more public current knowledge controversies in the environmental domain. Part of the controversy revolves around the Intergovernmental Panel for Climate Change (IPCC), which is mandated to assess the current knowledge about the state and causes of climate change. This case documents a particular episode in the history of the IPCC where it faced considerable scrutiny about its ways of working. The case illustrates the role of the linear model in these controversies (see Section 4.2) and the role of politics (Section 4.3).

Introduction

In the aftermath of the 15th Conference of the Parties to the UN Framework Convention on Climate Change, held in Copenhagen in December 2009, a media storm arose over the illegal publication of emails written by leading climate scientists ('climategate') and errors in the Intergovernmental Panel on Climate Change (IPCC) assessment report of Working Group II (see Pearce, 2010; Parry et al., 2007).[1] The heated nature of the controversies that followed indicates the extent to which the IPCC – not least as a result of being awarded the Nobel Prize in 2007 – has come under public scrutiny.[2]

This case explores features of public controversies with reference to events surrounding climategate and the discovery of errors in the 2007 IPCC report of Working Group II. The 'unsettlements' (Whatmore, 2009) surrounding these

---

[1] The IPCC was set up jointly by the World Meteorological Organization (WMO) and the United Nations Environment Programme (UNEP) in 1988. Its task is to undertake a comprehensive, objective, open, and transparent review of the status of research on non-natural global warming, its observed and projected impacts, and the policy response options available (options for adaptation and greenhouse gas reduction) (A/RES/43/53, 1988). For an overview of the genesis and early development of the IPCC, see Agrawala (1998). For its organisation and rules of procedure, see Petersen (2011).

[2] For a detailed reconstruction see Pearce (2010).

events offer a unique opportunity to exemplarily study public knowledge controversies on scientific evidence for policy making because 'back stage' practices inside the IPCC and climate science are displayed 'front of stage'. It is a case where experts are operating under the 'public microscope' (IAC, 2010). The case focuses on the controversy surrounding the making of climate expertise in its broader cultural context; in other words, it focuses not only on scientific facts or truth claims as such, but also on the public performance of science.

The Linear Model of Expertise

One of the major controversies in climate science focuses on the issues of whether climate change is occurring ('detection') and how much of this change can be attributed to human activities ('attribution') (Edwards and Schneider, 2001). The Fourth IPCC reports demonstrated that scientific evidence for global warming is now overwhelming, even if scientific projections of future climatic changes remain shrouded in uncertainty (Solomon et al., 2007). More than anything else, however, the Nobel Peace Prize awarded jointly to the IPCC and Al Gore in 2007 is also seen as an acknowledgement of the political impact of the former: by providing 'sound' scientific evidence, the panel has succeeded in simultaneously creating an awareness of the risks of climate change among both politicians and the wider public and in 'kick starting' the political activities required to address it. Leading IPCC representatives hold on to the concept of science as value free and its relative autonomy from politics, ideas deriving from the linear model of expertise.

The controversy over the publication of stolen emails demonstrates how these idealised notions of expertise make the IPCC vulnerable to the systematic amplification of doubt (Sarewitz, 2010; Wynne, 2010). As explained in Chapter 4, if, as the linear model holds, epistemic authority counts as the only political authority, decision making is reduced to a question of whether the science is right or wrong. As a result, evidence can easily be made to unravel.

First, the use of this idealised model of expertise leads to the 'schizophrenic' position of having an awareness of the political terrain while at the same time ignoring it (see also Pielke, 2007). While the IPCC claims to be detached from politics, it exercises a considerable amount of political influence and acts as a powerful agent in climate politics, as the Nobel Peace Price indicates. These prevalent idealised understandings of science increase public vulnerability to backlash campaigns. Lobby groups from the oil and car industries (mainly in the USA) realised the relevance of the IPCC's reports to their own business strategies, and by the end of the 1980s had formed a coalition consisting of representatives of the OPEC states and lobbyists working on behalf of US energy and automobile corporations. The emergence of this coalition shows that IPCC findings are received in a highly politicised context, where almost every scientific finding can have extensive implications for the stakeholders concerned. By characterising global warming as a major problem that requires multilateral and far-reaching political action to reduce greenhouse gas emissions, the IPCC's findings challenge vested interests, disrupt long-standing social relations, and question deeply ingrained lifestyles.

Second, this coalition, dubbed as consisting of *merchants of doubt* (Oreskes and Conway, 2010), tried to undermine the scientific evidence provided by the IPCC, for political and ideological reasons, in order to forestall mitigation action (Lewandowsky et al., 2015). The climate controversies also show that even when scientists, politicians, and publics agree on the basic principles such as the linear model of expertise, there is still plenty of room for disagreement about what the implications of that science are for action. This *doubt is our product* strategy contributed to the *politicisation* of the scientific debate on global warming. Rather than arguing about the political interests and values which motivated the controversy on climate change in the first place, all political actors became embroiled in a controversy over the scientific foundations on which their views were based and fought over the exclusive authority to interpret the findings of climate science. There is a strikingly *symmetrical form of asymmetry* here: both sides of the climate controversy – mainstream climate scientists, as represented by the IPCC, and its opponents alike – act as if climate policy will be decided by science alone (Pielke, 2007). They start from the premise that the 'united scientific voice' will 'trigger' political action. The idea is that once we agree what the science says, policy will automatically follow. Both sides in the controversy, including opponents of the IPCC, also insist that policy can't proceed until the science is sound. The controversy is caused by disagreements about the need for and urgency to set up mitigation policies.

Trust

Climategate turned out to be a question not only of the quality and integrity of climate science, but also of public confidence: 'Telling people "Hey, trust me –

I'm an expert" is just no longer enough' (Schiermeier, 2010, 170). These events demonstrate that trust in science does not depend on the strength of internal consensus among scientists alone, as many scientists may have hoped, given that the IPCC reports signalled that controversies over the existence of global warming have finally been settled. Indeed, many scientists have become frustrated that the evidence they have already provided has not prompted political action and invoked public trust.

Climategate fed into larger concerns about the reliability of the IPCC and the trust placed in it: could politicians and the public still rely on the IPCC to provide a sound assessment of scientific knowledge on climate change and still trust the IPCC's key messages (see PBL, 2010). The IPCC response shows that public trust in experts is related to the performance and persuasive power of the people and institutions who speak for science (Jasanoff, 2010; Hajer, 2012). The IPCC chairman, Dr Rajendra Kumar Pachauri, was reluctant to admit to errors such as overstating the rate of Himalayan glacial melting and positing a higher percentage of land lying below sea level in the Netherlands than is actually the case. Pachauri's comment suggested that the quality of the IPCC and its performance will triumph over doubts. As Pachauri put it: '[b]ased on the performance that we show to the whole world and the leadership that I provide to the IPCC, these opinions by a few motivated individuals will be washed away. I have no doubt about it at all' (Bagla, 2010, 510–511). It was only after a one-month delay – and under pressure – that the panel conceded that these minor errors had slipped through the system and had subsequently been corrected. The poor performance of the IPCC leadership in response to the Himalayan case exacerbated the problem of trust, and huge damage has been done to the reputation of climate science, although the scientific quality of the IPCC reports was never seriously challenged (Beck, 2012a).

When it comes to the illegal publication of climate scientists' emails, the 'gate' suffix was also used to link them to the disclosure of 'dirty doings' by the White House under US president Richard Nixon (Jasanoff, 2010). The media attention to the climategate affair focused on the statement in one hacked email that suggested scientists were willing to 'hide the decline' in observed global temperatures. Taken out of context, this statement could be seen to imply that scientists were colluding to deceive policy makers about observed temperature trends. This allegation is not borne out by deeper analysis of either the emails or the actions of scientists. Opponents, however, claimed that the IPCC authors had deliberately circumvented scientific review procedures and had falsified scientific results for political reasons to create a public mood of 'suspicion'. Exposing putative backstage corruption is common in controversies over science advice (e.g. Hilgartner, 2000). Even if the evaluations following climategate did not find evidence for serious corruption and conspiracy, they reveal a high degree of politicisation inside climate science (IAC, 2010; Latour, 2012). The publication of emails shows that some areas of climate science are plagued by intense antagonism. They also raised awkward questions about whether the IPCC was using peer review as a gatekeeping approach to defining what is sound information and what are appropriate and useful interpretations in the debates about climate science.

Pachauri's attempt to intimidate the critics effectively mimics the strategy of its opponents: in describing the IPCC's critics as 'politically motivated', Pachauri portrays the panel as untainted, distinct from politics' sound scientific expertise (Bagla, 2010). In pursuing the same strategy as his critics, however, Pachauri was merely adding fuel to the fire – polarising the debate and politicising climate change science. Science is thus reduced to a spectacle of assessments competing against one another for supremacy – a kind of 'contact sport', as Schneider (2009) has termed it. This trend is exaggerated by the tendency of the media to focus on duelling scientists and extreme, outlier opinions. The responses by the IPCC were based on the assumption that public doubts are caused by the politicisation of science by powerful corporate interests (Oreskes and Conway, 2010), the media's misrepresentation of dissent, and poorly informed public opinion.

As a result, mainstream climate scientists and their opponents alike present their scientific findings and associated policy agendas as true, beyond the need for deliberation. Claims of dissenters and opponents are countered by relativising them, highlighting their uncertainties, exposing their underlying political motivations, or pointing to their lack of authoritative backing from the mainstream scientific community (Elzinga, 1996). Thus, both sides of the knowledge controversy attack particular forms of knowledge production on the side of their opponents without relinquishing their commitment to science's cognitive autonomy and neutrality. In that way, they did not challenge, but rather reinforced the linear model of expertise (Beck, 2012b).

Cultural Divergence

However, climategate was more than just a dispute between two coalitions who were trying to shoot holes in the other coalition's science. There are remarkable cultural differences underlying the controversy over the IPCC.

Comparative studies show that disagreement expressed in public disputes about scientific evidence is often rooted in more fundamental differences over values or about the role of science in policy making (Jasanoff, 2011). These approaches help us to understand how and why knowledges are publicly challenged as legitimate claims to truth about the environment. National responses to climategate show that the demands for public accountability vary to a quite remarkable extent, and the IPCC enjoys different levels of credibility in different nation-states. A number of bloggers have made various demands regarding the transparency and accessibility of the Panel. Critics in some countries, including the USA, have bemoaned the fact that the IPCC has failed to fulfil specific stringent requirements of public accountability. These forms of politicisation are sparked by, among other things, the Panel's consensus procedures, the lack of a disclosure mechanism, and a perceived lack of legitimacy. These controversies reflect crucial political and normative disagreements caused by divergent national policy cultures in which knowledge-making is embedded. The IPCC's style of knowledge-making faced more resistance in the USA and India, for instance, than in Germany (see Jasanoff, 2011; Beck et al., 2016). Indeed, the Indian government expressed its concerns loudly whilst establishing alternative expert networks; many so-called climate change deniers in the USA and the UK used this event to promote their own causes; while in Germany the controversy failed to dent the broad political consensus on the urgency of climate action (Jasanoff, 2011; Mahony, 2014). These national variations cannot be explained solely in terms of quality of scientific knowledge, because it is the same body of knowledge (produced by the IPCC) that provides the common point of reference. The controversies also indicate potential problems with the IPCC's attempt to universalise one particular mode of knowledge-making and public reasoning (Hulme, 2013).

Thus, as Chapter 4 has also argued, this case shows that public trust in experts cannot be reduced to a function of information or fixed by more and better science. While the IPCC has been able to settle the scientific controversy on attribution and detection, the degree of uptake of its messages and its credibility in the eyes of citizens and policy makers around the world still varies significantly. The credibility of knowledge claims and trust in experts relates to long-established, culturally situated practices of interpretation and reasoning that function differently in different national and transnational contexts. This can partly explain why knowledge controversies like climategate flare up as they do.

References

Agrawala, S. (1998). Context and Early Origins of the Intergovernmental Panel on Climate Change. *Climatic Change*, 39(4), 605–620.

A/RES/43/53. (1988). Protection of Global Climate for Present and Future Generations of Mankind. www.un.org/documents/ga/res/43/a43r053.htm

Bagla, P. (2010). Climate Science Leader Rajendra Pachauri Confronts the Critics. *Science*, 327(5965), 510–511.

Beck, S. (2012a). Between Tribalism and Trust: The IPCC under the Public Microscope. *Nature and Culture*, 7(2), 151–173.

Beck, S. (2012b). The Challenges of Building Cosmopolitan Climate Expertise: The Case of Germany. *WIRES Climate Change*, 3(1), 1–17.

Beck, S., Forsyth, T., Kohler, P. M., Lahsen, M., and Mahony, M. (2016). The Making of Global Environmental Science and Politics. In U. Felt, R. Fouché, C. A. Miller, and L. Smith-Doerr, eds., *The Handbook of Science and Technology Studies* (pp. 1059–1086). Cambridge: MIT Press.

Edwards, P. N., and Schneider, S. H. (2001). Self-governance and Peer Review in Science for Policy: The Case of the IPCC Second Assessment Report. In C. Miller and P. N. Edwards, eds., *Changing the Atmosphere: Expert Knowledge and Environmental Governance* (pp. 219–246). Cambridge: MIT Press.

Elzinga, A. (1996). Shaping Worldwide Consensus – The Orchestration of Global Climate Change Research. In A. Elzinga and C. Landström, eds., *Internationalism and Science* (pp. 223–253). London: Taylor Graham.

Hajer, M. (2012). Living the Winter of Discontent: Reflections of a Deliberative Practitioner. In M. Heinlein, C. Kropp, J. Neumer, A. Poferl and R. Römhild, eds., *Futures of Modernity: Challenges for Cosmopolitical Thought and Practice* (pp. 77–94). Bielefeld: Transcript.

Hilgartner, S. (2000). *Science on Stage: Expert Advice as Public Drama*. Stanford: Stanford University Press.

Hulme, M. (2013). *Exploring Climate Change through Science and in Society: An Anthology of Mike Hulme's Essays, Interviews and Speeches*. London: Routledge.

IAC, (2010). *Climate Change Assessments: Review of the Processes and Procedures of the IPCC*. Amsterdam. http://reviewipcc.interacademycouncil.net.

Jasanoff, S. (2010). Testing Time for Climate Science. *Science*, 328(5979), 695–696.

Jasanoff, S. (2011), Cosmopolitan Knowledge: Climate Science and Global Civic Epistemology. In J. Dryzek, R.B. Norgaard, and D. Schlosberg, eds., *Oxford Handbook of Climate Change and Society* (pp. 129–143). Oxford: Oxford University Press.

Latour, B. (2012). Reflexive Modernity Brings Us Back to Earth. In M. Heinlein, C. Kropp, J. Neumer, A. Poferl and R. Römhild, eds., *Futures of Modernity: Challenges for Cosmopolitical Thought and Practice* (pp. 65–75). Bielefeld: Transcript.

Lewandowsky, S., Oreskes, N., Risbey, J., Newell, B., and Smithson, M. (2015). Seepage: Climate Change Denial and its Effect on the Scientific Community. *Global Environmental Change*, 33, 1–13.

Mahony, M. (2014). The Predictive State: Science, Territory and the Future of the Indian Climate. *Social Studies of Science*, 44(1), 109–133.

Oreskes, N., and Conway, E. (2010). *Merchants of Doubt: How a Handful of Scientists Obscured the Truth on Issues from Tobacco Smoke to Global Warming*. New York: Bloomsbury Press.

Parry, M., Canziani, O., Palutikof, J., Van der Linden, P., and Hanson, C., eds. (2007). *Climate Change 2007: Impacts, Adaptation and Vulnerability. Contribution of Working Group II to the Fourth Assessment Report of the Intergovernmental Panel on Climate Change*. Cambridge: Cambridge University Press.

PBL (2010). *Assessing an IPCC Assessment: An Analysis of Statements on Projected Regional Impacts in the 2007 Report*. Bilthoven: PBL (Netherlands Environmental Assessment Agency).

Pearce, F. (2010). *The Climate Files: The Battle for the Truth about Global Warming*. London: Guardian Books.

Petersen, A.C. (2011). Climate Simulation, Uncertainty, and Policy Advice: The Case of the IPCC. In G. Gramelsberger and J. Feichter, eds., *Climate Change and Policy: The Calculability of Climate Change and the Challenge of Uncertainty*. Dordrecht: Springer, 91–111.

Pielke, R. (2007). *The Honest Broker: Making Sense of Science in Policy and Politics*. Cambridge: Cambridge University Press.

Sarewitz, D. (2010). World View: Curing Climate Backlash. *Nature*, 464(7285), 28.

Schiermeier, Q. (2010). Few Fishy Facts Found in Climate Report. *Nature*, 466(7303), 170.

Schneider, S. H. (2009). *Science as a Contact Sport: Inside the Battle to Save Earth's Climate*. Washington: National Geographic.

Solomon, S., Qin, D., Manning, M., Chen, Z., et al, eds., (2007). *Climate Change 2007: The Physical Science Basis. Contribution of Working Group I to the Fourth Assessment Report of the Intergovernmental Panel on Climate Change*. Cambridge: Cambridge University Press.

Whatmore, S. J. (2009). Mapping Knowledge Controversies: Science, Democracy and the Redistribution of Expertise. *Progress in Human Geography*, 33(5), 587–598.

Wynne, B. (2010). When Doubt Becomes a Weapon. *Nature*, 466(7305), 441–442.

Case C Whose Deficit Anyway? Institutional Misunderstanding of Fracking-Sceptical Publics

LAURENCE WILLIAMS AND PHIL MACNAGHTEN

*Hydraulic fracturing, or fracking, is a relatively new technique used to extract oil or natural gas. It is a highly contested technique, with often-fierce disagreements about its safety and environmental impacts between experts, governments, oil companies, and inhabitants. This case focuses on public resistance to fracking, offering a useful illustration of the limitations of the information deficit model (*Section 4.4*) and showing that controversies cannot be resolved by assuming that opposing actors suffer from an information deficit that can be rectified by means of better communication and greater public understanding of science.*

Hydraulic Fracturing (or Not) in the UK – A 'Knowledge Controversy'?

Hydraulic fracturing is a technique used in the exploitation of oil and gas resources, particularly 'unconventional' hydrocarbons such as shale gas. Hydraulic fracturing induces cracks in underground rock layers by injecting high-pressure fluids into perforated, sometimes horizontal, wells. This is required to stimulate gas particles trapped in the pores of shales in order for the gas to flow to the surface. The more commonly used abbreviation 'fracking' is often employed as a catch-all term to refer to the entire process of exploiting shale gas – from site preparation and construction, to well abandonment and site restoration, and the ultimate use of the resource.

Fracking has enjoyed high levels of support and enthusiasm from UK governments since 2010, particularly from within the Conservative Party who formed, initially, the senior partner in a governing coalition, and since 2015, a majority government.[1] It has also enjoyed strong support from engineers and geologists, who, notwithstanding some notable dissenting voices, have

---

[1] That is, till 2017. What follows primarily concerns the English fracking controversy. Distinct dynamics are found to an extent in Wales, and especially Scotland and Northern Ireland.

tended to mirror the support and enthusiasm of Westminster. Other experts, perhaps most notably in energy systems and public health, have on balance tended to be somewhat more sceptical.

Fracking has been highly controversial in the UK, much as elsewhere.[2] The exploitation of shale gas has been promoted by supporters as having strong potential as a growth-promoting, secure, and sustainable energy source.[3] However, fracking operations – actual and proposed – have given rise to grassroots protest groups across the country. Environmental concerns have been raised about possible contamination of water resources, high levels of water consumption, waste disposal, induced seismicity, and public health. Additionally, there are wider concerns about the industrialisation of rural landscapes, the reconcilability with climate change commitments, the exaggeration of benefits, industry–academic relations, democratic deficits in policy and decision making, and more besides.

Key early moments in the UK controversy included two small seismic events caused by the company Cuadrilla when fracking in the north-west of England in 2011, as well as 'scare stories' emanating from the United States, most notably from the infamous documentary *Gasland* (Fox, 2010). The UK government imposed a de facto moratorium on fracking after those seismic events, while the Royal Society and the Royal Academy of Engineering conducted a joint report into the risks of fracking (The Royal Society and The Royal Academy of Engineering, 2012). The report concluded that the risks of fracking could be effectively managed so long as operational best practices were ensured through regulation, and these findings were used by the UK government to justify giving the 'green light' to new operations. Expert assurances of manageable risk, however, appear to have had little success in reassuring and placating significant proportions of the public who remain at best sceptical, with a well-organised minority fully committed to, and mobilised in, opposing fracking.

Public controversy has been most conspicuously evident in direct action protests in the West Sussex village of Balcombe in the summer of 2013, alongside campaigns by local groups in Lancashire, and more recently North Yorkshire, including efforts to resist developments through the planning system. Beyond these highly involved, 'campaigning publics',[4] polling has

---

[2] Shale gas resources are present on every continent, with controversy typically accompanying development wherever it is proposed. See Wood (2012) and Steger and Milicevic (2014) for summaries of controversy globally.

[3] See, for example, Department for Communities and Local Government and Department of Energy and Climate Change (DCLG/DECC), 2015.

[4] Mohr et al. (2013) helpfully distinguish between four different types of 'publics', namely 'campaigning', 'latent', 'civil society', and 'diffuse'.

regularly shown ambivalence in the broader public's attitudes towards fracking and the exploitation of shale gas (see, for example, DECC, 2016). This 'diffuse public' has arguably been slowly but gradually turning against fracking since the Balcombe protests in 2013 (O'Hara et al., 2015).

Questions over why fracking has proved controversial, whether it deserves to be, and how the situation might be resolved have been characterised by the diagnosis of various deficits, identifying blame and responsibility in what various actors are said to lack (i.e. deficits), and offering various attendant solutions.

Who Lacks What? (And What to Do About It?)

As was pointed out in Chapter 4, knowledge controversies often feature some version of the deficit model (it may make sense to talk of deficit models in the plural). A deficit model purports to explain controversy over a new technology or domain of science, and apportion blame for that controversy, by highlighting a particular kind of deficit or lack on the part of a particular group. Appropriate solutions for resolving this controversy, then, depend on *who is said to lack what*. Perhaps the most commonly attributed deficit said to cause controversy is that of a public knowledge or information deficit (the information deficit model; see Chapter 4). In this version of the deficit model, it is the public's lack of knowledge that explains and is to blame for the existence of controversy. Controversy will therefore be resolved by addressing this public deficit or lack, and so the solution entails educating the public.

A knowledge deficit is not the only controversy-causing lack that the public might typically be diagnosed with. Other commonly attributed public deficits in so-called 'knowledge controversies' include those of 'attitude' and 'trust' (Wynne, 2006; Bauer et al., 2007). Deficits in 'attitude' are claimed in cases where publics fail to recognise the necessity of new innovation or when they are not sufficiently 'in love with science' (Bauer et al., 2007). Deficits in 'trust' involve unwillingness on the part of the public to defer to remote authority, or a tendency to suspect. Alongside pedagogical efforts, deficits in attitude and trust are seen by some as best addressed through various kinds of public participation or engagement, ranging from transparency and 'one-way' interactions to more thorough, deep, and dialogic modes of public involvement.

These *public* deficit models, and the scientific and political elites that typically wield them, have been heavily criticised by scholars in Science and Technology Studies (STS). Public knowledge deficits, of course, are acknowledged to exist, but these are often insufficient in themselves to explain controversy (Wynne, 2006). First, the public knowledge deficit model is criticised on the grounds of ambiguous evidence about the relationship between public knowledge of a controversial technology and public attitudes to it. For example, survey research rarely reports emphatic and unambiguous support for the underlying assumption 'The more you know, the more you love it' (Bauer et al., 2007).[5] This suggests that public ignorance may not be the driver of controversy, and that increasing public knowledge – the information deficit model's go-to solution – may not necessarily or straightforwardly resolve controversy. Second, the public attitude deficit model is criticised for being general and indiscriminate. That is, it mistakenly assumes that publics possess a singular, homogeneous attitude to an equally homogenous 'science'. In practice, publics' attitudes to different areas and applications of science are typically more discriminate, complex, and discerning than a blanket favourable or unfavourable attitude to science as a whole (Wynne, 1991). Finally, the public trust deficit model is criticised for failing to recognise that alternatives to delegative modes of knowledge and decision making exist.[6] Furthermore, proposed solutions to public trust deficits in the form of public engagement often neglect questions of framing and purpose or fall into the category of 'sales pitch', and hence have been viewed historically as inherently 'untrustworthy' attempts to instrumentally secure support under the (false) guise of 'public participation' (Wynne, 2006).

Building on this critique, STS scholars have put forward a series of *institutional* deficit models, as alternative causes of and explanations for controversy over science and technology to the *public* deficits already considered. These deficit models shift the focus to various shortcomings of scientific and political elites and institutions. According to these institutional deficit models, controversy can be explained by the institutional tendency to underemphasise uncertainty and leave contestable assumptions or commitments underexamined (Wynne, 1996; Yearley, 2005). Furthermore, institutions tend to misunderstand sceptical publics and miscategorise their concerns. They reduce and ignore complex and varied concerns about values, purposes, ethics, inclusivity, and so on by framing the issue as one of 'risk and safety' (Macnaghten and Carro-Ripaldo, 2016; Williams et al., 2017), thereby exacerbating controversy through a failure to recognise and respond to public concerns. Institutional deficit models also point to an institutional lack of self-awareness, self-

---

[5] Bauer et al. (2007) suggest that 'empirical investigations of the knowledge/attitude relationship have remained inconclusive until recently', with surveys showing 'a small positive correlation between knowledge and positive attitudes' (p. 84). There is, furthermore, a larger variance amongst the knowledgeable and on controversial issues (Allum et al., 2008).

[6] See Callon et al. (2009) for one such alternative, which they term 'dialogic'.

reflection, and self-criticism. Specifically, they suggest that science, and political actors accessing the authority of science, fail to recognise the contingency of their own knowledge, the exclusionary and depoliticising consequences of reducing public issues to scientific ones, the double bind of dependency and mistrust this places publics in, and, finally, the insult that is added to injury when this situation is blamed on various public deficits (Wynne, 1996, 2007). Solutions to institutional deficits generally centre on reforming these institutions, their cultures, and their behaviour towards greater humility, openness, and reflexivity.

The fault line, then, between the attribution of deficits in knowledge controversies falls between emphasising public or institutional shortcomings in the apportioning of blame for the development and sustenance of a controversy. Are these controversies caused by 'infantile publics' who are in need of education and engagement by institutions, or by 'blinkered institutions' who should focus their efforts on reforming themselves? These two models – the infantile public and the blinkered institution – are obviously coarse, and, as with any nationwide science-based controversy lasting more than half a decade, elements of each are likely to ring true from time to time. But we argue nevertheless that they can be used as heuristic models that highlight the different deficits and dynamics argued to drive controversy. We will now turn to the UK fracking case in more detail in order to test these models as explanations for controversy.

The Fracking Case: Infantile Publics or Blinkered Institutions?

In this section we explore accusations of public knowledge deficits on the one hand (from the 'infantile public' model), and institutional reflexivity and understanding deficits on the other (from the 'blinkered institutions' model), as putative explanations for the fracking controversy, before concluding in favour of the latter as the more plausible account.

Explaining the level of public opposition to fracking through appeals to 'myths' and 'scaremongering', a scientifically illiterate media, and irresponsible non-governmental organisations (NGOs) are all regular features of UK institutional rhetoric on fracking. The then Prime Minister David Cameron, for example, spoke of 'myths' in a 2013 op-ed in the Sunday Telegraph entitled 'We cannot afford to miss out on shale gas' (Cameron, 2013); the then Energy Minister Andrea Leadsom referred to 'fracking scaremongering' (quoted in

Schofield, 2016); Cuadrilla's chief executive Francis Egan has voiced his anger at 'scaremongering' opponents (quoted in Gosden, 2015); and Ken Cronin, the chief executive of the onshore oil and gas operator's industry body, has spoken of 'unfounded myths' (quoted in Shale Gas International, 2015).

Supplying publics with 'the facts about fracking', moreover, has been an oft-proffered solution to the controversy. Averil MacDonald, for instance, the chair of UK Onshore Oil and Gas, has suggested:

> The science behind fracking is well understood: it's been used for more than 60 years. But we need to use a clear, accurate application of scientific evidence to help reassure local communities that reserves of British natural gas can be developed safely and with the minimum of environmental impact. (Macdonald, 2015, no pagination)

A lack of public recognition of the necessity of fracking, the resources it could unlock, and the broader benefits it could bring have been a key part of institutional rhetoric. Andrea Leadsom, for example, has suggested that the national need for gas, 'for years to come', and the apparent economic and environmental benefits of shale gas, are an 'inconvenient truth' for the 'anti-fracking lobby', who 'don't want to acknowledge the economic and environmental benefits shale gas could bring' (Leadsom, 2015). More generally, the fracking controversy, according to MacDonald and many others, is characterised by a neat distinction between 'scientific facts' on one side, and 'emotional fears' on the other (MacDonald, 2015). The institutional rhetoric presented here conforms to the 'infantile public' model. The existence of controversy over fracking is explained by and blamed on public deficits of various kinds, particularly knowledge deficits, and envisaged solutions involve communicating what is presented as settled, authoritative knowledge to sceptical publics.

Is there any justification for these explanations of and solutions for the fracking controversy? Evidence supporting the 'infantile public' model would include survey research suggesting a positive association between public knowledge of fracking and public support for fracking. Further evidence would involve public engagement and information provision initiatives being found to secure public support through education. The evidence from UK survey research on the relationship between public knowledge of and support for fracking is, in fact, mixed and somewhat ambiguous.

Two long-running surveys have tracked public attitudes to fracking in the UK. The Department for Business, Energy and Industrial Strategy (BEIS)[7] conducts an Energy and Climate Change Public Attitudes Tracker involving

---

[7] The tracker was formerly run by the Department of Energy and Climate Change (DECC). The DECC was incorporated into the new department BEIS on 14 July 2016.

questions on shale gas and fracking,[8] and the University of Nottingham in
collaboration with Yougov has tracked attitudes to shale gas extraction since
March 2012.[9]

Stedman et al. (2016) report that, in general, 44 per cent of respondents think
shale gas extraction should be allowed ('support'), 27 per cent think that it
should not be allowed ('opposition'), and 29 per cent indicate that they don't
know. Levels of support amongst those who answer the knowledge question
correctly (who we'll call 'the knowledgeable') and those who don't (who we'll
call 'the non-knowledgeable') are presented in Table C1. These findings
suggest that the information deficit model appears to stand up reasonably
well here, as the knowledgeable are twice as likely to support extraction than
the non-knowledgeable. Of course, opposition rises slightly with knowledge
too, though the infantile public model would presumably suspect these knowledgeable opponents of having been duped by unreliable or erroneous information. However, we know that support among the knowledgeable dropped to
46.5 per cent (with opposition at 36.1 per cent) by September 2015 (O'Hara et
al., 2015), and again to 37.3 per cent (with opposition at 41.1 per cent) in
October 2016 (O'Hara et al., 2016). Support among the knowledgeable therefore appears to be on a downward trend, but no analysis has yet been published
on how this change has affected the snapshot presented in Stedman et al.
(2016).

The BEIS survey muddies the water further for the 'infantile public' model.
It asks respondents to self-assess how much they know about fracking for shale
gas (with four options: 'a lot', 'a little', 'aware of it but don't really know what
it is', and 'never heard of it') and whether they support or oppose it (with six
options: strongly support, support, neither support nor oppose, oppose, strongly
oppose, and don't know). Table C1 presents the attitudes to fracking for shale
gas amongst those who either know a lot or a little (roughly equivalent to
Stedman et al.'s 'knowledgeable' category) and those who've heard of it but
don't really know what it is along with those who have never heard of it (i.e.
who might be expected to not know the answer to Stedman et al.'s knowledge
question). Attitudes have been aggregated to total support (support + strong

[8] Wave 8 (conducted December 2013) was the first to ask both a knowledge and an attitude
question. At the time of writing, the most recently published results are from wave 18 (conducted
June/July 2016). The publicly available dataset from wave 9 is incomplete and so not involved in
our analysis, leaving a total of ten waves (see BEIS, 2016 for a summary).

[9] The Nottingham survey uses a knowledge question to measure respondent knowledge. Although
there have been ten rounds to date, those who failed to answer the knowledge question correctly
exited the survey up until September 2014. This means that only the three most recent rounds
enable an analysis of the relationship between knowledge and attitude, of which September 2014
is the only round from which analysis on this relationship has so far been published (Stedman et
al., 2016).

Table C1: *BEIS tracker*[10] *compared to Stedman et al. (2016)*

Wave of BEIS tracker	8	10	11[11]	12	13	14	15	16	17	18	Stedman et al. (2016)
Attitude amongst those who say they know a lot or a little about fracking (roughly equivalent to 'the knowledgeable')											'The Knowledgeable'
Total Support	37%	32.5%	33.7%	31.8%	33.1%	28.8%	30.4%	29.6%	25.2%	28.1%	50%
Total Opposition	30.7%	34%	35.6%	35.2%	37.7%	39.9%	41.9%	42.2%	44.2%	41%	29%
Don't Know	32.3%	33.6%	30.6%	33.1%	29.1%	31.3%	27.7%	28.8%	30.6%	30.9%	20%
Attitude amongst those who say they are aware of it but don't really know what it is and those who have never heard of it (roughly equivalent to 'the non-knowledgeable')											'The Non-knowledgeable'
Total Support	17.8%	14.6%	16.6%	15.5%	15.4%	12.8%	15.1%	15.9%	14%	12.2%	25%
Total Opposition	9.9%	12.7%	13%	8.4%	10.9%	14.1%	16.0%	10.6%	15.4%	15.1%	22%
Don't Know	72.3%	72.7%	70.4%	76.1%	73.7%	73.1%	68.9%	73.5%	70.6%	72.6%	53%

[10] The authors' work based on DECC/BEIS datasets available here: www.gov.uk/government/collections/public-attitudes-tracking-survey.

[11] Wave 11 was conducted at roughly the same time as the Nottingham September 2014 survey on which the Stedman et al. (2016) analysis is based.

support), net opposition (opposition + strong opposition) and don't know (neither support nor oppose + don't know) to make the results as comparable as possible with the previous survey. The Stedman et al. (2016) results are included to aide comparison.

The BEIS findings are a challenge to the infantile public model. From wave 10 onwards, 'the knowledgeable' have been more likely to oppose than support fracking, and levels of support and opposition are regularly comparable amongst 'the non-knowledgeable', with 'don't know' always by far the most common response for this group. As the Department of Energy and Climate Change (DECC) (as they were then) themselves put it in the Wave 16 summary: 'Support for fracking appears to be inversely linked to awareness, as those who know more about fracking tend to be more likely to oppose it' (DECC, 2016, 6).

We could speculate that the differences between these two surveys are down to online versus face-to-face interview data-collection techniques, slightly different sampling strategies, different ranges of options for attitudes, or different ways of measuring 'knowledge'. However, the point for now is that survey research on the relationship between knowledge and attitudes in the UK is inconclusive and ambiguous, and may indeed be shifting towards sceptical attitudes to fracking amongst the 'knowledgeable' (however measured). It certainly falls short of justifying the emphatic adoption of the infantile public model in the institutional rhetoric presented above. Furthermore, that tweaks in method can produce such large differences across two surveys serves to high-light the difficulties in reliably measuring public knowledge, attitudes, and the relationship between them. A final point worth raising is that when institutional actors refer to 'scaremongers' it is unlikely that they mean the 'diffuse publics' captured through survey research. They're more likely referring to 'campaign-ing' publics actively involved in opposing fracking. This latter group would almost certainly, in its entirety, fall into Stedman et al.'s (2016) 'knowledge-able' category (i.e. they would recognise the definition of shale gas). Of course, whilst the ambiguity reported above troubles the 'infantile public' model, it also fails to provide positive support for the 'blinkered institutions' model, which maintains that institutional misunderstanding of and behaviour towards publics also drive controversy.

Across the large range of activities that might come under the banner of 'UK public engagement on fracking', there is very little publicly available assess-ment on whether these processes are indeed improving public knowledge of fracking and securing public support for it in doing so. There is certainly little discernible effect of these efforts on the 'diffuse publics' captured through survey research in the above discussion. There has been one formal government-backed process, though, in which Sciencewise and TNS BMRB[12] conducted a series of public dialogues for DECC, designed to learn lessons on how to 'effectively' engage the public on fracking. The TNS report, which assesses the dialogues, acknowledges that most participants began the dialogues with low familiarity and a relatively open mind about fracking. Furthermore, whilst some participants remained neutral or became positive as they 'learned', 'many were prone to become more negative as the dialogue progressed' (TNS BMRB, 2014, 14). 'This was a trend', they continue, 'that became stronger as the workshops progressed and participants learnt more, which will be vital to address when designing engagement activities' (TNS BMRB, 2014, 14). They suggest the reasons why the provision of information failed to secure support in these cases included confirmation bias, perceived unknowns, the complexity of the issue, and a lack of confidence in decision-making bodies.

Taken at face value, these reasons challenge the 'infantile public' model's envisaged solution, as clearly information provision does not necessarily or straightforwardly succeed in securing support. However, they also reveal a degree of institutional blinkeredness. The institutions involved framed continued public scepticism as a problem to be solved through the design of engagement processes. Instead of treating expressed concerns as potentially valid and public engagement as an opportunity to genuinely recognise, register, and respond to them, they treated public concern as error (confirmation bias) and perception (of uncertainty, complexity, and decision makers) to be corrected through the design of engagement processes. Instead of a lack of confidence in decision-making bodies giving rise to considerations about how to reform decision making to inspire greater public confidence, they considered it a barrier to securing support to be addressed in the design of further engagement activities. In this way, the explanation of and blame for continued scepticism is placed entirely on publics (their errors and perceptions), leaving institutional knowledge claims, assumptions, commitments, and behaviour shielded from scrutiny. This is precisely the type of institutional behaviour referred to by Brian Wynne when he identifies 'inherently' untrustworthy attempts to instrumentally secure support under the (false) guise of public participation (Wynne, 2006). This lack of reflexivity, inability or unwillingness to respond to public concern, and overlooking of the contestability of institutional claims and commitments conform to the 'blinkered institutions' model.

---

[12] Sciencewise is the UK's national centre for public dialogue in decision making involving science and technology; TNS BMRB is a UK social research agency.

Further evidence supporting the 'blinkered institutions' model comes
from work exploring public perceptions of fracking using focus groups
conducted by the authors in 2013 with largely latent, uninvolved publics.[13]
Four themes of public concern were found that went considerably beyond
the technical 'risk and safety' framing adopted by the government. First, the
trustworthiness of institutional actors was questioned by participants,
including the motivations suspected to be behind emerging institutional
pro-shale commitments, and the credibility of promised benefits, particularly
the likelihood of their equitable distribution. Second, the importance of
inclusive decision making was stressed amidst scepticism about genuine
public influence on policy and concern over the perceived likelihood of the
capture of policy-making processes by powerful incumbent interests. What
was viewed as an under-considered drift towards fracking in policy, suspected to be *unresponsive* to contrary views, was a third concern. Finally,
participants tended to be highly *precautionary* when encountering complexity, uncertainty, and contingency to do with fracking or its governance. That
this preference for a precautionary approach was perceived to be in sharp
contrast to prevailing attitudes to incertitude in decision- and policy-making
institutions was a further cause for concern. These concerns, it should be
noted, are not primarily related to areas of specialised knowledge (e.g. risk),
but are in the main suspicions and expectations about institutional
behaviour,[14] including the expected capacity and willingness to register
these and other public concerns – in other words, institutional blinkeredness. As such, they are not principally erroneous epistemic beliefs that
could be straightforwardly corrected with the provision of authoritatively
compelling knowledge (if and where such knowledge was available), as the
infantile public model envisages. Instead, addressing them would look first
and foremost to institutional practices and culture, as highlighted by the
blinkered institutions model. Furthermore, this brings into question the
status of the UK fracking controversy as a 'knowledge controversy'.
Although public concern about fracking is often closely related to, and
involves, specialist knowledge, it overflows the boundaries of the category
'knowledge' as conventionally understood. As such, explaining controversy
by and blaming it on a public knowledge deficit, as the 'infantile public'
model widely adopted by UK institutional actors does, is at best partial, if
not misplaced.

---

[13] Reported in more detail in Williams et al. (2017).
[14] The final preference for precaution and concern that this is not shared by institutions and elites
comes closest, but is 'meta-epistemic' rather than epistemic. That is, it concerns a preference for
a style of response to incertitude, rather than claims about the contents of knowledge directly.

Conclusion

In conclusion, the above case provides tentative support for the 'blinkered institutions' model – and not the infantile public model – as the heuristic that best explains the dynamics of controversy over fracking in the UK. First, the survey evidence is ambiguous and certainly falls some way short of the emphatic support that would be required to justify the enthusiastic adoption of the 'infantile public' model in the institutional rhetoric cited earlier. There are, furthermore, signs of a trend towards more sceptical and oppositional attitudes both generally, and amongst the knowledgeable (however measured). This constitutes a serious challenge for the 'infantile public' model: ignorance does not appear a necessary condition for opposition. Second, there is little evidence to support the view that information-provision-oriented public engagement secures support for fracking by reducing supposedly culpable public knowledge deficits. The assessment, discussed above, of the Sciencewise/TNS BMRB dialogues suggests that providing publics with information does not necessarily or straightforwardly have the effect of securing their support as if it were a simple dose–response relationship. This problematises the 'infantile public' model's go-to solution for resolving controversies. Furthermore, the assessment also reveals the involved institutions' framing of participant scepticism as an undesired response to uncritically presumed (by those institutions) compelling information, to be designed out of future 'effective' engagement processes. This framing aligns entirely with the blinkered institutional account. Public engagement emerges here as a cynical sales pitch and post hoc decision justification, as opposed to a sincere attempt at democratic renewal and institutional learning. What is to be changed through these processes are sceptical public perceptions and attitudes, leaving the government commitment in favour of fracking shielded from, and unresponsive to, criticism. Third, and finally, further support for the 'blinkered institutions' model comes from the focus group evidence reported above. A specialist knowledge based 'risk and safety' framing of the fracking issue is unable to recognise, register, and respond to clearly articulated broader concerns and criticisms about, say, inclusivity and trustworthiness, which are emphatically more to do with institutional behaviour than erroneous beliefs concerning settled knowledge. Taken together, a picture is painted here of institutions diagnosing public deficits as causing controversy with insufficient evidence, and pursuing attendant solutions premised on at least an partial misunderstanding of public responses. In doing so, they render themselves insensitive and unresponsive to key drivers of public concern in ways that are likely to exacerbate rather than resolve the controversy over fracking in the UK.

References

Allum, N., Sturgis, P., Tabourazi, D., and Brunton-Smith, I. (2008). Science Knowledge and Attitudes Across Cultures: A Meta-analysis. *Public Understanding of Science*, 17 (1), 35–54.

Bauer, M. W., Allum, N., and Miller, S. (2007). What Can We Learn from 25 Years of PUS Survey Research? Liberating and Expanding the Agenda. *Public Understanding of Science*, 16(1), 79–95.

BEIS (2016). *BEIS Public Attitudes Tracker – Wave 18 Summary Tables*. Department for Business, Energy and Industrial Strategy, London. www.gov.uk/government/statistics/public-attitudes-tracking-survey-wave-18

Callon, M., Lascoumes, P., and Barthe, Y. (2009). *Acting in an Uncertain World: An Essay on Technical Democracy*. Cambridge: MIT Press.

Cameron, D. (2013). We Cannot Afford to Miss out on Shale Gas. *The Telegraph*, 11 August. www.telegraph.co.uk/news/politics/10236664/We-cannot-afford-to-miss-out-on-shale-gas.html

DCLG and DECC (2015). *Shale Gas and Oil Policy Statement by DECC and DCLG*. Department for Communities and Local Government and Department of Energy and Climate Change, London. www.gov.uk/government/publications/shale-gas-and-oil-policy-statement-by-decc-and-dclg/shale-gas-and-oil-policy-statement-by-decc-and-dclg

DECC (2016). *DECC Public Attitudes Tracker – Wave 16*. Department of Energy and Climate Change, London. www.gov.uk/government/uploads/system/uploads/attachment_data/file/602460/PAT_Summary_Wave_16__2_.pdf

Fox, J. (2010). *Gasland*. United States: New Video Group.

Gosden, E. (2015). Cuadrilla Chief Francis Egan: Scaremongering Fracking Opponents Make Me Angry. *The Telegraph*, 17 January. www.telegraph.co.uk/finance/newsbysector/energy/oilandgas/11352111/Cuadrilla-chief-Francis-Egan-Scaremongering-fracking-opponents-make-me-angry.html

Leadsom, A. (2015). *Shale Gas – An Inconvenient Truth for the Anti-fracking Lobby*. https://decc.blog.gov.uk/2015/09/23/shale-gas-an-inconvenient-truth-for-the-anti-fracking-lobby/

Macdonald, A. (2015). It's True Women Don't like Fracking. I Want to Change That. *The Guardian*, 23 October. www.theguardian.com/commentisfree/2015/oct/23/fracking-shale-gas-women

Macnaghten, P., and Carro-Ripaldo, S. (2016). *Governing Agricultural Sustainability: Global Lessons from GM crops*. London: Routledge.

Mohr, A., Raman, S., and Gibbs, B. (2013). *Which Publics? When? Exploring the Policy of Involving Different Publics in Dialogue around Science and Technology*. London: Sciencewise.

O'Hara, S., Humphrey, M., Andersson-Hudson, J., and Knight, W. (2015). *Public Perceptions of Shale Gas Extraction in the UK: Two Years on from the Balcombe Protests*. Nottingham: University of Nottingham.

O'Hara, S., Humphrey, M., Andersson-Hudson, J., and Knight, W. (2016). *Public Perceptions of Shale Gas in the UK: From Positive to Negative*. Nottingham: University of Nottingham.

Schofield, K. (2016). *Andrea Leadsom: 'Don't Be Browbeaten by Fracking Scaremongering'*, Politicshome. www.politicshome.com/news/uk/energy/house/73121/andrea-leadsom-dont-be-browbeaten-fracking-scaremongering

Shale Gas International. (2015). *Dispelling Shale Gas Myths – Guest Interview.* 13 July. www.shalegas.international/2015/07/13/dispelling-shale-myths-guest-interview/

Stedman, R. C., Evensen, D., O'Hara, S., and Humphrey, M. (2016). Comparing the Relationship between Knowledge and Support for Hydraulic Fracturing between Residents of the United States and the United Kingdom. *Energy Research & Social Science*, 20, 142–148.

Steger, T., and Milicevic, M. (2014). One Global Movement, Many Local Voices: Discourse(s) of the Global Anti-fracking Movement. In L. Leonard and S. B. Kedzior eds., *Occupy the Earth: Global Environmental Movements* (pp. 1–35). Bingley: Emerald Group Publishing Limited.

The Royal Society and The Royal Academy of Engineering. (2012). *Shale Gas Extraction in the UK: A Review of Hydraulic Fracturing*. London. https://royalsociety.org/~/media/policy/projects/shale-gas-extraction/2012–06–28-shale-gas.pdf

TNS BMRB. (2014). *Public Engagement with Shale Gas and Oil: A Report on Findings from Public Dialogue Workshops*. London. www.gov.uk/government/uploads/system/uploads/attachment_data/file/382345/Public_engagement_with_shale_gas_and_oil__TNSBMRB__Final_for_publication.pdf

Williams, L., Macnaghten, P., Davies, R., and Curtis, S. (2017). Framing 'Fracking': Exploring Public Perceptions of Hydraulic Fracturing in the United Kingdom. *Public Understanding of Science*, 26(1), 89–104.

Wood, J. (2012). *The Global Anti-Fracking Movement: What it Wants, How it Operates and What's Next*. London: Control Risks.

Wynne, B. (1991). Knowledges in Context. *Science, Technology, & Human Values*, 16 (1), 111–121.

Wynne, B. (1996). May the Sheep Safely Graze? A Reflexive View of the Expert–Lay Knowledge Divide. In S. Lash, and B. Wynne eds., *Risk, Environment and Modernity, Towards a New Ecology* (pp. 44–83). London, Thousand Oaks, New Delhi: SAGE Publications.

Wynne, B. (2006). Public Engagement as a Means of Restoring Public Trust in Science – Hitting the Notes, but Missing the Music? *Community Genetics*, 9(3), 211–20.

Wynne, B. (2007). Public Participation in Science and Technology: Performing and Obscuring a Political–Conceptual Category Mistake. *East Asian Science, Technology and Society: An International Journal*, 1(1), 99–110.

Yearley, S. (2005). *Making Sense of Science: Understanding the Social Study of Science*. London: SAGE Publications.

5

The Limits to Knowledge

WILLEMIJN TUINSTRA, AD RAGAS, AND WILLEM HALFFMAN

5.1 Introduction

This chapter deals with the limits to our knowledge. In spite of our high expectations of science for the solution of environmental problems, scientific results are fringed with uncertainties. Measurement equipment always has limited precision, even though it is continuously improving; shifting oil prices make it difficult to assess the cost-effectiveness of alternative energy sources; unpredictable human behaviour might off-set potential ingenious technical solutions; and so on. In this chapter, we show the differences between risk and uncertainty and provide you with the conceptual tools to distinguish different types of risk and uncertainty. We also describe the consequences of uncertainty for environmental policy, and of the social response to uncertainty in environmental knowledge. Ultimately, this will provide you with the tools to recognise forms of uncertainty, to recognise typical reactions, and to understand strategies and debates addressing uncertainty and risk.

5.2 Different Conceptions of Uncertainty

First, we have to define what we mean by uncertainty. One way of looking at uncertainty is especially significant in the context of decision making because it relates to conflicting interpretations of reality. This is the way social scientists normally tend to look at uncertainties, while it is of as much importance to natural scientists because of its close relation to problem definition and the choice of system boundaries. As we have seen in the chapter on frames (Chapter 3) and the previous chapter on controversies (Chapter 4), various interpretations of an environmental problem may exist in society *and* in science, and different views on solutions may compete. The same problem can be framed in different ways, and thus analysed using different scales, measures, conceptual models, and time frames. This also implies variations in possible regulatory frameworks. In fact, this is true for all societal problems, but it becomes especially visible in cases such as climate change, preventing loss of biodiversity, genetically modified organisms, urbanisation, and water management. Due to their complexity, the lack of knowledge, and the existence of many groups with various stakes in the issue, it turns out to be especially difficult to come up with one unanimous definitive analysis of a problem. Thus, because of the existence of *multiple frames* of a problem, any chosen representation of the problem will inherently be associated with uncertainty. Different audiences will judge this uncertainty differently. In addition, because each representation will highlight different aspects of the problem, each representation will also highlight different uncertainties.

If you have a natural science background, you will generally tend to view uncertainty mainly in the light of *limited scientific* knowledge: our representation of the world is incomplete, but improving. The causes of limited knowledge can be manifold, including inexact measurements, incomplete data, and impossibility to measure, as well as to more fundamental limits of what can actually be known. In this view, uncertainty exists when the representation of the world does not match the 'real' situation of the world. Hence doing research is a means of acquiring more knowledge and reducing uncertainty – that is, new knowledge will diminish the gap between the representation of the world and the real situation. Many scientists will equate reducing 'uncertainty' with reducing the size of their confidence intervals (statistical uncertainty). We want to know how sure we are, and we try to calculate this. However, it is important to note that there is a difference between reducing confidence intervals and diminishing the gap between representation and reality. Acquiring new knowledge might lead us to recognise that we have to calculate our confidence intervals differently, which means they can actually *increase*. For example, ongoing climate research reveals complexities in the climate system and new uncertainties, which increase the uncertainty we are already aware of (Van der Sluijs, 2005). In the same way, increased sampling frequency could lead to more data and hence more knowledge, but these data could then enable us to discern new patterns that contradict the initially inferred causal relationships. While we initially were certain about expected patterns and processes, we then are less certain that we really understand the system. In this case we have the counterintuitive result that more research leads to less scientific certainty (van Asselt and Vos, 2006; Shackle, 1955), at least if we equate uncertainty with confidence intervals. We are less sure than we were before. In other words, we realise that there had been some unknown unknowns that we had not taken account of in our previous calculations. On the other hand, thanks to our new knowledge we still might feel we have diminished the gap between representation and 'reality'. Many people will equate this with reduced uncertainty. It is important to be aware of the existence of these different conceptions of uncertainty, as they are used interchangeably and can give rise to misunderstandings.

Above, we have put the word 'reality' in quotation marks for a reason. Next to frame diversity and limited knowledge, the difficulty of representing the 'real' state of the environment has another cause. Biological and social systems are characterised by *variability*, due to randomness, variation, chance, or chaotic behaviour in various processes. This variability is a given, which research might help to better understand and describe by uncovering particular patterns, but which cannot in itself be 'reduced'.

Before proceeding to discuss the relations between uncertainty, risk, and decision making, let us first wrap up our examination of various conceptions of uncertainty by way of an example. Say we are interested in whether a pesticide harms human health. Under laboratory conditions we can test whether the farmer experiences any harmful effects by inhalation when he applies the prescribed doses to the crop. We limit scientific uncertainty by using highly sensitive measuring equipment, selecting a sophisticated sampling method, and doing careful calculations. It might turn out that we do not find any harmful effects and that we are quite sure about this, as confidence intervals turn out to be small. However, there might still be a gap between knowledge and reality, because we only tested under laboratory conditions. We might suspect that, in the routine use of pesticides, users actually apply higher doses than recommended and do not bother with protection measures. In order to test this, we will also monitor the actual behaviour of the users in the field (e.g. observation, questionnaires) and we will take blood samples. Based on our findings a decision could be made to prescribe certain protecting clothes. We might find that the pesticide can be safely used under these conditions. However, protective clothing will not protect the neighbours from drifting pesticides. We realise that we have to increase our system boundaries and take blood samples from the neighbours as well. However, by increasing the system boundaries beyond the laboratory and the workers in the field, the researchers have fewer factors under control, and in addition have to deal with variability in weather conditions and unpredictable behaviour of both the workers and the neighbours. Therefore, while the researchers have increased their knowledge and insight in terms of the specific relevant processes, they might find they are actually less confident about what they can say with certainty about the effects of the pesticide on human health.

Basically, we see two simultaneous but different strategies to develop knowledge dealing with environmental issues. On the one hand, while trying to grasp the complexity of the issue, we increase the complexity of its representation in order to get a picture as complete as possible. On the other hand, in an attempt to grasp the issue in order to support decision making, we inevitably have to simplify the system and search for a way to get to the essentials. This 'getting to the essentials' implies making choices, as the previous chapter on framing has shown. These choices depend on the purpose of our knowledge development (*why*), which will determine *what* we decide to be relevant to describe or not, and *how* we do that (Kovacic and Giampietro, 2015). In our pesticide example, initially only the health of the farmers was in focus. This determined what we started measuring and monitoring. If we are concerned about the wider health impacts of the pesticide, we include a wider range of people who could possibly be exposed. However, we could have chosen to look not only into pesticide drifting, but also into run-off of toxins into the water after heavy rainfall, or the impacts of the pesticide across a larger area, or on ecosystems instead of human health.

What follows is that the different conceptions of uncertainty in our representation of 'reality' are very much interconnected, yet they look at uncertainty from different perspectives and do not per se complement each other. The chosen frame will influence the way we judge whether our knowledge is limited or not and whether we have to improve our methodology, our measuring, or our means of describing relationships in the system. In the same way, it will influence which variability in the system we see as significant. This in turn will have an influence on how we judge risk, as we will see in the next section.

5.3 Different Conceptions of Risk

What is risk? The lemma in the Cambridge Dictionary gives as the first option 'the possibility of something bad happening'. In the Oxford English Dictionary, we find risk to be 'a situation involving exposure to danger'. Traditionally, scientists have defined risk as a function of the effect or the impact of a particular event and the probability this event will happen. Thus Risk = Effect × Probability. Some people will take 'exposure' as an additional factor, but others will say this is part of calculating the effect or the probability. Calculating risk makes sense when, for example, a decision has to be made about evacuating a region because of a hurricane. When meteorological experts warn that there is a high probability that the hurricane will hit the area, in most cases the authorities will decide to evacuate. Calculation of risk will also be of help for these same authorities in less stressful times, for example to decide on what kind of measures could be taken to protect the population against future hurricanes. If, for example, all houses can be made hurricane-proof, next time evacuation might not be necessary, because the effects of the hurricane can be supposed to be less severe, even when the probability of a hit would be the same.

However, in the case of environmental and health issues (or hurricanes, for that matter), more often than not neither effects nor probabilities of occurrence are easily determined. This uncertainty is due to limited knowledge as well as to the inherent variability or indeterminacy of the system.

Another problematic feature of the traditional conception of risk lies in the limited criteria to decide on risk populations and time horizons. Of all the people that might possibly be affected, now or by knock-on effects in the future, who will be considered? This too depends on our framing of the issue. Meanwhile, we have to decide whether to consider collective effects on populations, or effects for specific groups. The probability of a hurricane in our region might be known, but we still do not know whether our village or our house will be hit. In addition, the determination of effect, damage, or loss supposes a shared understanding of what counts as damage or what loss is relevant. If authorities are concerned about the more vulnerable part of the population, they might also be more inclined to account for this group in risk calculations. Framing again is crucial here. This is all the more relevant for decision makers, because risk perceptions will vary with judgements on relevant damage.

If decision makers' ideas about relevant damage or relevant populations differ too much from those in significant parts of society, then this will lead to distrust. Because trust is crucial in decision making concerning environmental and health risks, before turning to decision makers' strategies to deal with risks in practice, we will first discuss the relation between risk perception and trust.

5.4 Risk Perception and Trust

It is important to acknowledge the psychological background of risk perception. Public acceptance of risk is influenced by more than just the number of critically affected people. For the acceptance and valuation of risk, it makes a difference whether the potential danger is due to natural causes or activities of certain actors. People will perceive earthquakes differently if they are caused by tectonic drift than if they result from mining activities. Furthermore, people are less tolerant of risk when the effects are expected to be irreversible. Distribution effects play a role as well. If certain groups of people are more affected than others, they will perceive this as unfair and be less tolerant of the risk, such as with the unequal effects of Hurricane Katrina. Whether risks are calculated and presented as a collective or an individual risk therefore makes a difference for how people will perceive the risk and authorities' actions in respect of that risk. Furthermore, people tend to weigh voluntary and involuntary risks differently (Slovic, 2000, 2010). Thus, even though it might follow from model calculations that a risk is identical for all people involved, how the risk will be valued by different people will vary widely depending on who is affected, how, and at whose responsibility. For example, as we have seen above, risk perceptions will vary with judgements on relevant damage.

Transparency about assumptions and what is taken to be relevant in risk calculation is therefore important; otherwise, misunderstandings may occur which can easily give rise to distrust. The way in which authorities communicate about risk thus affects trust (Löfsted, 2005). On the one hand, lack of communication can have a negative impact on trust. On the other hand, communication strategies that are based on a misinterpretation of reasons for (dis)trust may have the unintended effect of increasing distrust (Löfstedt, 2005; Chapter 4, this volume). For example, focusing too much on the safety of certain technological solutions, while neglecting the fear of misuse of accompanying rules, might give the impression of deliberate distraction from important issues. Another issue is that the degree of perceived uncertainty will influence risk perceptions. Knowledge about certain problems is distributed differently in society, and access to knowledge and trust in experts will influence the way people will assess risk. As also becomes clear from the case on flower-bulb farming (Case D: Angry Bulbs), lack of transparency with the authorities about what is known or why this knowledge is not more precise might increase fear and feelings of unsafety and distrust.

Communication, perceived uncertainty, risk perception, and trust are thus interconnected processes (Löfstedt, 2005). In situations where the determination of risk is not straightforward, it therefore makes sense to follow more interactive and deliberative strategies in deciding which risks matter. Decision makers simply have to deal with risk and uncertainty in practice, as they cannot be resolved entirely by relying on scientific experts alone. Which strategies they follow and which roles can be played by science is the topic of the next section.

5.5 Dealing with Uncertainty in Policy Practices

In policy practice, the question of 'who decides whether it matters at all?' is an important one. These choices are not choices that can be made by 'science' or determined by 'research'. In cases of risks that cannot be easily calculated, dimensions of time, relevant populations, acceptable damage, and manageable costs or procedures have to be weighed against each other, which is a matter of choice, perspective, and politics, rather than a matter of calculating solutions. However, what we often see in these cases is that in order to escape difficult choices or for other reasons, policy makers resort to experts for conclusive evidence and definite answers (Van Asselt and Vos, 2006; Chapters 3 and 4, this volume). In general, society tends to expect clear answers from science. Also, groups of concerned citizens and NGOs start looking for scientific evidence in the expectation this will bring the solution, in particular in cases of high uncertainty. Van Asselt and Vos (2005, 2006) coin the phrase 'uncertainty paradox' as an umbrella term for situations in which uncertainty is present and acknowledged, but the role of science is framed as one of providing certainty.

In practice, this is not possible for science precisely because risk calculations require value choices and framing. As we have argued, the decision of what to protect and what protection level is acceptable is not something scientific experts can decide upon on their own. In addition, there exists another level of choice which involves judgements of experts and which also inevitably implies value choices. What to include or exclude in calculations depends on judgements by experts on what is important or, as seen earlier, on what is feasible given the necessity of reduction of complexity in representation. Often, in practice these choices are resolved by somewhat arbitrary or negotiated standards and norms. For example, health risks posed by air pollution are calculated in terms of Disability-Adjusted Life Year (DALY), i.e. loss of a year of healthy life. What these years actually mean for the individual is not specified. In the Netherlands, in defining acceptable flood risks, norms for collective risks and individual risks differ, by an arbitrary factor, to compensate for differences in risk perception. In other countries, the distinction might not be made at all. These kinds of choices are inevitable, and in a way necessary, because we want to base our decisions on something. However, to present them as a scientific necessity generates misunderstandings.

Other strategies followed in policy practice make uncertainty a more explicit part of deliberation. An example is the use of the *precautionary principle*. The precautionary principle is a maxim for environmental policy stating that 'where there are threats of serious or irreversible damage, lack of full scientific certainty shall not be used as a reason for postponing cost-effective measures to prevent environmental degradation' (Rio declaration, Principle 15). The precautionary principle might be used as 'an overarching framework of thinking that governs the use of foresight in situations characterised by uncertainty and ignorance and where there are potentially large costs to both regulatory action and inaction' (Harremoes et al., 2001). In this way, the precautionary principle legitimates decisions and actions in situations characterised by uncertainty. It is generally agreed that uncertainty is the essence of the precautionary principle (Van Asselt and Vos, 2006). The precautionary principle is thus a principle that is specifically applicable in dealing with risks that are not easy to calculate. However, paradoxically there are intrinsic difficulties in its applicability in this very case. Van Asselt and Vos (2006) note that while the regulation invoking the precautionary principle acknowledges the limits of science in providing conclusive evidence, it still requires scientific decisiveness on whether something constitutes a risk, and for this reason would justify the applicability of the precautionary principle. At the same time, in these cases we do have some sense that there is a threat of 'serious or irreversible damage'. Therefore, as noted earlier, instead of requesting scientific certainty in such a case, it is more productive to weigh various dimensions of the problem against each other in interaction with those who are affected.

Scientific research can play different roles here. On the one hand, it can contribute to opening up the framing of a problem and offer new insights and perspectives within the public debate. This is, for example, the purpose of the study into the exposure of local residents to pesticides described in the bulb-farming case (Case D: Angry Bulbs). Instead of calculating risk with conventional exposure scenarios, samples are taken to test exposure in the field, which shifts the framing of the problem from theoretical application of pesticides towards application in the field. On the other hand, science might also narrow down perspectives, as described in the case on carbon capture and storage (CCS) in Chapter 9 (Case J: Groupthink and Whistle Blowers in CO_2 Capture and Storage). Furthermore, the public debate can limit the room for scientific research. As we have seen in Chapter 3 on frames, if a certain frame is dominant in public discourse, the room for doing scientific research might be limited to this frame, thus reinforcing it.

As authors such as Stirling and Gee (2002) show, a precautionary approach asks for acknowledgement of the limits of knowledge, some kind of humility with regard to what we can know. It supposes an extended scope to wide areas of possible harm and, next to theoretic models and laboratory tests, requires attention to monitoring and research, just like in our pesticide example. For this reason, a precautionary approach requires the participation of a full range of interested and affected parties. As an approach for regulation purposes, Stirling and Gee (2002) stress the importance of addressing the pros and cons of various policy options, and of technology strategies that are both flexible and robust enough to be less vulnerable to surprises and unknown unknowns. Adaptive management strategies could be a way forward.

5.6 Conceptualising Strategies to Deal with Uncertainty and Risk

Broadly speaking, the three sources for uncertainty identified at the start of this chapter – multiple possible frames, variability and indeterminacy, and limited knowledge – also require three varying strategies to deal with them (Brugnach et al., 2008). Multiple frames imply 'knowing too differently' and *addressing the relation between multiple frames and actors*. Limited knowledge implies 'knowing too little' and *remedying the deficiencies in the available knowledge*. Variability implies 'accepting not knowing' and seeking ways to *manage the system with its inherent variability and indeterminacy*. Obviously, in practice, uncertainties associated with environment and health issues do not root in just one of the discerned sources and will require various strategies at the same time.

A different connection between uncertainty and various management strategies has been made by Klinke and Renn (2006), relating them to risks. What options do you have and what strategies can you follow in order to deal with uncertainty in the case of possible danger? The strategies are based on risk classes, which in an exercise with the German Scientific Advisory Council for Global Environmental Change (WBGU, 2000) were given names of Greek mythological figures. The idea is that identifying risk classes can help political decision makers to select management strategies. The risk classes are distinguished by probability of occurrence (high or low) and the extent of damage (high or low) in combination with a judgement about the certainty of the former two. Certainty here means scientific certainty, therefore probabilities or extent of damage are labelled as 'uncertain' in cases of limited knowledge or when relevant processes are characterised by variability and indeterminacy. Klinke and Renn distinguish three central categories of risk management, namely science-based strategies, precautionary strategies, and more deliberative strategies. The two risk classes *Damocles* and *Cyclops* require mainly science-based management strategies, taking as a starting point that limited knowledge is not the problem. For the risk classes *Pythia* and *Pandora*, probabilities and

Table 5.1: *Overview of risk classes and management strategies (based on Klinke and Renn, 2006)*

Risk Class (*with example*)	Extent of damage	Probability of occurrence	Management	Strategies for action
Damocles(nuclear energy)	High	Low	Science-based	Reducing disaster potential
				Ascertaining probability
Cyclops (earthquakes)	High	Uncertain		Increasing resilience
				Preventing surprises
				Emergency management
Pythia(genetic engineering)	Uncertain	Uncertain	Precautionary	Implement precautionary principle
Pandora(initial use of CFCs)	Uncertain	Uncertain		Developing substitutes
				Improving knowledge
				Reduction and containment
				Emergency management
Cassandra(climate change)	High	High	Discursive	Consciousness building
				Confidence building
Medusa(electro-magnetic fields)	Low	Low		Public participation
				Risk communication
				Contingency management

extent of damage are uncertain and application of the precautionary principle is required. Partly, uncertainty can be remedied by increasing the knowledge base, but also the system with its inherent variability or indeterminacy has to be managed. For the risk classes *Cassandra* and *Medusa*, probability and extent of damage are supposed to be (scientifically) certain because experts agree, but in society other frames exist and different values are found relevant, thus discursive and deliberative strategies involving various actors are needed for building consciousness, trust, and credibility (Klinke and Renn, 2006).

1. ***Damocles.*** Damocles had to sit at a banquet while a razor-sharp sword was suspended above him on a thin thread. A fatal event could occur for Damocles any time even if the probability is low. Hence, in this risk class, low probability is combined with a high extent of damage. Typical examples mentioned by Klinke and Renn are technological risks associated with nuclear energy, large-scale chemical facilities, and dams.
2. ***Cyclops.*** This giant with one eye personifies risks of which we only can have a limited view. Possible damages are known, but the causes or relevant processes cannot be overseen. Probabilities cannot be determined, but if

anything goes wrong the damage can be considerable. Natural hazards such as earthquakes and floods belong to this class, but also e.g. infectious diseases.

3. *Pythia.* This seer of the oracle of Delphi personifies risks of which both the probability of occurrence and the extent of damage remain uncertain. Some relations between events are assumed, but Pythia's predictions are always ambiguous. Examples are the use of genetic engineering in agriculture or self-enforcing global warming.

4. *Pandora.* Pandora's box seemed harmless as long as it was not opened. When it was, it turned out to contain evil gifts. Unknown unknowns are at issue here. Certain activities can seemingly cause no harm, and the extent of damage done is only revealed years later. An example is the use of CFCs, which only after many years were identified to cause ozone depletion. In the case of Pandora's box, probabilities of occurrence and the extent of damage are unknown, and we have no idea of causal relationships.

5. *Cassandra.* Nobody believed this princess, who predicted the Trojan War and the defeat of Troy. Cassandra personifies risks for which the probability and impact are known to be considerable, but which various actors tend to downplay. For example, because the timescales on which such damage occurs are large, people tend to take the damage less seriously. Klinke and Renn mention biodiversity loss and climate change as examples.

6. *Medusa.* Medusa could frighten people even in her absence. She personifies risks for which experts judge the probability, and the known damage, as small and manageable, while society perceives them as considerable. Thus, there is a discrepancy between scientific risk assessment and risk perception in society. Electromagnetic fields are an example wherein most experts do not find significant epidemiologically or toxicologically significant adverse effects, while many people feel involuntarily affected.

Obviously, classifying the various risks is not straightforward. The classes are not given by nature. Of course, risks for which the probabilities and extent of damage are known and which can be managed on a scientific base do exist. But what makes experts or decision makers decide that probabilities and extent of damage are known? As we have seen, this depends on choices of relevant populations and what counts as damage. It also depends on how much tolerance towards risks decision makers can permit themselves given the tolerance towards the risk in society. So, classes of risk are themselves something to make choices about: those involved in developing the management strategy will have to decide what kind of risk they will treat a given issue as. It is not only the Medusa and Cassandra cases that require deliberation and interaction with a full range of interested and affected parties to sort out what is important and what should be taken into account. All risk classes ask for deliberative methods, acknowledgement of diversity, and weighing of different views, if only to identify them.

5.7 Conclusion

In this chapter, we introduced two ways of conceptualising uncertainty, which in practice tend to be used interchangeably: (1) uncertainty expressing the gap between how the world works and how we can represent this, (2) uncertainty expressing our judgement of that gap. In the first case, new knowledge can *reduce* uncertainty. In the second case, this might happen as well, but new knowledge could also *increase* our sense of uncertainty, changing our judgement of the gap. The way people form this judgement depends not just on knowledge, but also on their frame and values as well as on their awareness of other frames and values. We also showed how frames determine how people judge risk. Assessing risks, as well as determining which risks are relevant, implies making choices and value judgements. Therefore, this is not just a matter of science but also of political choice. Policy strategies like the precautionary principle and adaptive management strategies make uncertainty an explicit part of deliberation. A precautionary approach asks for acknowledgement of the limits of knowledge, requiring participation of affected parties to include a broader knowledge base and to address the pros and cons of various policy options. Adaptive management asks for monitoring, learning, and adapting, requiring dialogue and participation in a similar way.

References

Asselt, M. B. A. van, and Vos, E. (2005). The Precautionary Principle in Times of Intermingled Uncertainty and Risk: Some Regulatory Complexities. *Water, Science and Technology*, 52(6), 35–41.

Asselt M. B. A. van, and Vos, E. (2006). The Precautionary Principle and the Uncertainty Paradox. *Journal of Risk Research*, 9(4), 313–336.

Asselt M. B. A. van, and Vos, E. (2008). Wrestling with Uncertain Risks: EU Regulation of GMOs and the Uncertainty Paradox. *Journal of Risk Research*, 11(1), 281–300.

Brugnach, M., Dewulf, A., Pahl-Wostl, C., and Taillieu, T. (2008). Toward a Relational Concept of Uncertainty: About Knowing Too Llittle, Knowing Too Differently, and Accepting Not to Know. *Ecology and Society*, 13 (2), 30. www.ecologyandsociety.org/vol13/iss2/art30/

Harremoes, P., Gee, D., MacGarvin, M., et al. (2001). *Late Lessons from Early Warnings: The Precautionary Principle 1896–2000*. Copenhagen: European Environment Agency.

Klinke, A., and Renn, O. (2006). Systemic Risks as Challenge for Policy Making in Risk Governance. *Forum Qualitative Sozialforschung*, 7(1), 33. http://nbn-resolving.de/urn:nbn:de:0114-fqs0601330

Kovacic, Z., and Giampietro, M. (2015). Beyond 'Beyond GDP Indicators': The Need for Reflexivity in Science for Governance. *Ecological Complexity*, 21, 53–61.

Löfstedt, R. E. (2005). *Risk Management in Post-trust Societies*. Hampshire/New York: Palgrave Macmillan.

Shackle, G. L. S. (1955). *Uncertainty in Economics and Other Reflections*. Cambridge: Cambridge University Press.

Slovic, P. (2000). *The Perception of Risk*. London: Earthscan.

Slovic, P. (2010). *The Feeling of Risk. New Perspectives on Risk Perception*. London: Earthscan.

Sluijs J. P. van der. (2005). Uncertainty as a Monster in the Science–Policy Interface: Four Coping Strategies. *Water Science and Technology*, 52(6), 87–92.

Stirling, A., and Gee, D. (2002). Science, Precaution, and Practice. *Public Health Reports*, 117, 521–533.

WBGU, German Scientific Advisory Council on Global Change. (2000). *World in Transition. Strategies for Managing Global Environmental Risks. Annual Report 1998*. Berlin: Springer.

Case D Angry Bulbs

AD RAGAS AND MARGA JACOBS

This case study of lily bulb farming and the effects experienced by local residents demonstrates the role uncertainty can play in environmental issues. It is an example of a complex situation in which stakeholders resort to experts for conclusive evidence and definite answers, yet these can rarely be provided. The case illustrates that in such situations with various problem definitions, limited knowledge, and variability at multiple levels, creating room for deliberation between stakeholders is a more productive strategy to create knowledge for action. Trust and acknowledgement of each other's interests are key for such deliberative processes.

Lily Bulb Farming

Bulb farming is an important economic activity in the Netherlands. In bulb farming, a considerable amount of pesticides is used to grow flowers. The use is highest in the farming of lily bulbs: 134.6 kg per hectare for lily bulbs versus 27.5 kg for tulip bulbs (data over the year 2012[1]).

The growth cycle of lily bulbs starts with decontamination of the soil to prevent the infection of young bulbs by root nematodes, the so-called 'wet soil decontamination'. Young lily bulbs are then planted, typically in March/April when the soil is sufficiently warm and dry. In the growing season, which generally lasts 4–5 months, pesticides are sprayed on a weekly basis. The bulbs are harvested between August and Christmas by removing the topsoil along with the bulbs. The bulbs are then flushed with water to remove the soil. The reclaimed soil is put back on the land. In order to minimise the risk of pests, the fields on which lily bulbs are grown are rotated on a regular basis.

---

[1] Data from Dutch Statistics Office: Centraal Bureau voor de Statistiek, 2018. *Gebruik gewasbeschermingsmiddelen in de landbouw; gewas en toepassing* (application of pesticides); consulted on 11 September 2018: https://opendata.cbs.nl/statline/#/CBS/nl/dataset/84007NED/table?ts=1536674915462.

Lily Bulb Farming in Westerveld

Bulb farming traditionally takes place in the western part of the Netherlands. However, in the late 1990s, the government of the Zuid-Holland province imposed stricter environmental regulations on bulb farming, particularly on the farming of lily bulbs. The lily bulb farmers started looking for alternative locations, and spread over the country. One of the areas where bulb farming took off was Westerveld: a rural municipality with approximately 19,000 inhabitants, located in the eastern part of the Netherlands. Westerveld measures 28,300 hectares, including several small villages, agricultural land, and two national parks. Since 2001, the area covered by lily bulbs has varied between 113 hectares (2009) and 269 hectares (2012). Westerveld does not even have the largest area of lily bulb farming of the municipalities in the Netherlands. In 2014, the top five was led by their neighbour Midden-Drenthe (701 ha), followed by Noordoostpolder (516 ha), Hardenberg (398 ha), Hollandse Kroon (379 ha), and Schagen (313 ha).

Angry Bulb Foundation

In 2002, a number of citizens of Westerveld formed the Angry Bulb Foundation. These people were worried about the increasing area of land used for lily bulb farming. They initially focused on landscape deterioration, i.e. the lowering and smoothening of characteristic raised farm land, reflecting the historical use of fertilisers consisting of piled organic materials. Furthermore, bulb faming was associated with several other environmental threats, i.e. potential public health effects due to the intensive and potentially illegal use of pesticides, eutrophication effects due to the use of fertilisers, dehydration of natural areas due to drainage, and local disturbance effects due to frequent pesticide applications, local flushing installations, and increased traffic intensity in the harvest season.

At the time of writing, the Angry Bulb Foundation still exists. Their main aim is to end the nuisance caused by lily bulb farming. The group is an officially registered foundation with a governing board. The core group consists of three members, supplemented by a fluctuating membership of other people. The group has a website (www.bollenboos.nl) and makes newsletters.

In the first years of their existence, the Angry Bulb Foundation targeted local and regional authorities in order to persuade them to be more restrictive in respect of lily bulb farming. Several flushing sites were closed after the Angry Bulb Foundation argued that these sites were located in valuable landscapes.

The foundation also successfully appealed against a proposed new flushing site. In 2009, the municipality of Westerveld banned all flushing sites in open fields, allowing them only in built-up areas around the bulb farms – a great success for the Angry Bulb Foundation. However, the foundation was initially less successful in reducing pesticide use. The foundation proposed the introduction of a warning system for pesticide spraying to inform local residents living close to bulb fields, a pesticide-free zone next to roads, and measures to reduce the exposure of cattle. However, the foundation found out that local and regional authorities have only a limited mandate to restrict the use of pesticides because this is regulated at the national level due to European guidelines. Regional authorities did start an initiative for more sustainable bulb farming, but the authorities also made clear that legal action against pesticide drift would be unfeasible as long as there was insufficient scientific evidence of adverse health effects.

Risk Assessment for Pesticides

Before a pesticide product is allowed on the European market, potential risks for human health (i.e., operators, workers, bystanders, and consumers), animal health, and the environment (aquatic ecosystems, terrestrial ecosystems, and wildlife species exposed after accumulation in food chains) must be assessed and demonstrated to be acceptable. Key components of such an assessment are an exposure scenario resulting in an exposure estimate, and a hazard assessment resulting in a reference value, which reflects the maximum exposure level that is considered safe given the available information. The ratio between the estimated exposure and the reference exposure is used as a proxy for risk, i.e. this ratio should be below 1.0. Risks are assessed for each endpoint separately (operators, workers, bystanders, consumers, animal health, aquatic ecosystems, terrestrial ecosystems, and wildlife), and, where relevant, a distinction is made between risks from acute exposures (high exposure for a short period of time) and chronic exposures (low exposures for a long period of time). In order to reduce costs, risks are generally assessed using a tiered approach. This means that risks are initially assessed conservatively, i.e. using few data and high safety factors. More refined and costly assessments are performed only when lower tiers show a potential risk (i.e. a risk ratio above 1.0).

The exposure scenarios considered in risk assessments are based on the assumption that the pesticide is applied in accordance with the prescription. Furthermore, conservative assumptions are typically applied to deal with variability in exposures and uncertainty in data. Reference values for human health, animal health, and ecosystems are typically derived from results of toxicity experiments with species such as rats, fish, daphnids (water fleas), and algae. Typically, the lowest exposure concentration showing no effect or a small effect (e.g. 5–10 per cent increase over the control group) is extrapolated to humans, animals, or ecosystems by applying safety factors to account for the uncertainties involved in this extrapolation. For example, the standard safety factor used to derive a safe exposure value from rats for humans is 100; often rationalised as a factor of 10 to account for potential differences in sensitivity between species (i.e. rats and humans) and another factor of 10 because the tested rat population is small and shows much less diversity in terms of composition, habits (food intake, use of medicines, sleeping pattern), and environmental conditions (temperature, light, humidity) than the human population.

National News

An important turning point in the discussion on the safety of pesticide use was the production and broadcasting of two documentaries by a national investigative news programme in the Netherlands. The first documentary, *Poison in Bulb Farming*, was broadcast in October 2010. It highlighted the worries of people living close to bulb fields about the intensive use of pesticides and potential health impacts, particularly on young children. Besides local people, several experts expressed their concern because the chronic exposure of people living close to bulb fields could in theory be considerable, e.g. by spray drift into their gardens and houses. However, this exposure had never been systematically analysed and thus was not taken into account in the risk assessment procedure for pesticides. There was also concern about the fact that multiple substances are used in practice, but that risks are assessed only for individual substances without considering potential mixture effects. The director of the Dutch Pesticide Assessment Agency was interviewed and admitted that these aspects were not covered in the assessment procedure. The documentary led to several questions being raised in the Dutch parliament. In response, the incumbent Secretary of State asked the Heath Council of the Netherlands, an independent scientific advisory board of the government on matters of health, for advice.

While the Health Council was working on its advice, a second documentary, *Lilies with a Funny Smell*, was broadcast on Dutch national television in November 2013. The documentary highlighted several health incidents related to the application of metam sodium (dithiocarbamate), an organosulfur compound used for soil decontamination. Metam sodium is injected into the soil where it slowly fumigates, killing most root nematodes. Under certain environmental conditions (e.g. fog and wind) the fumes may escape the soil and disperse into the local environment. The documentary interviewed several residents living next to lily bulb fields who suffered from health complaints (e.g. dizziness, fainting, and skin irritation) after the application of metam sodium. Pictures of dead fish were shown, and it was suggested that this was a result of the water being polluted with metam sodium. It was also noted that an attempt to introduce a 50-metre zone between lily bulb fields and vulnerable locations in the municipality of Hardenberg (e.g. schools and houses) did not gain sufficient support from local politicians due to a lobbying campaign set up by lily bulb farmers.

In May 2014, the Dutch Pesticide Assessment Agency decided to temporarily ban the use of metam sodium for soil decontamination because the probability of 'unacceptable risk' was considered too high. However, the farmers successfully appealed, and the Court of Appeal for Businesses overturned the decision in February 2018, arguing that a ban is disproportionate considering the risks involved. The risks of using metam sodium are currently being reassessed by the Dutch Pesticide Assessment Agency.

New Research

The Health Council of the Netherlands published its advice to the Secretary of State, called 'Crop protection and residents', in January 2014. In this report, the Health Council stated that the 'there is sufficient reason to initiate an exposure study among this section of the population [local residents] here in the Netherlands, and to adapt the approval procedure for plant protection products'. The advice also contained a number of suggestions to reduce the exposure of local residents. Based on this advice, the government commissioned a €10 million study into the exposure of local residents to pesticides and the potential health effects. The study is ongoing, and consists of an exposure part and a health part. The exposure part focuses on determining the application and dispersal of pesticides into potential exposure media for local residents, such as air, home dust, soil, and plants. Another important aspect is the determination of pesticides and metabolites in the urine of local residents. It is also investigated whether behavioural characteristics play a role in the exposure. The health part focuses on the relationship between registered health complaints and the proximity to certain crop areas such as lily bulb fields. This epidemiological effect study has an explorative character and does not involve experiments. Initially, the Health Council proposed undertaking this effect study only if the exposure study had shown elevated exposure levels for local residents, but the Secretary of State decided otherwise. There are two commit-tees advising the project: one stakeholder committee, and one scientific com-mittee. Important choices made at the start of the project included:

1. Which pesticides to include?
2. Which locations to include?
3. Should the illegal use of pesticides be included?

The latter question addresses an important concern of local residents, but could not be included in the study since farmers will not reveal when and where illegal use takes place. However, it is possible to analyse the samples taken for the presence of illegal pesticides.

At the time of writing, the study into the exposure of local residents to pesticides and potential health effects is ongoing. The results of the study are expected in late 2018 or early 2019.

Uncertainties

In this section, we will use the Angry Bulbs case to illustrate different types of uncertainty that play a role in environmental issues.

Framing Uncertainty

The first question is 'What is the problem?' Although the question is simple, the answer is not. Initially, a limited number of residents were worried about the increasing area of agricultural land used for lily bulb farming because they associated this with different types of environmental threats, of which land-scape deterioration was the most important one. But as time passed, the discussion increasingly focused on the potential health impacts of pesticide applications on local residents. At the same time, the number of stakeholders involved substantially increased. Initially only local residents, lily bulb farm-ers, and local and regional authorities were involved, but later journalists, scientists, and national authorities also contributed. All these stakeholders are likely to define the problem differently. For example, a scientist may define the problem as 'Can the 8 pesticides selected in the study be traced in the blood of local residents 24 hours to 1 week after application following prescribed procedures?' This definition of the problem is very narrow compared to the worries of the local residents that triggered the Angry Bulb Foundation. The lily bulb farmers may yet define the problem completely differently, e.g. 'A small number of people threatening their livelihood'.

What people define as a problem strongly depends on their perception of risk, and is influenced by factors such as values, cultural background, available knowledge, familiarity, control, voluntariness, and benefits. A lily bulb farmer is less likely to perceive high risks because the farmer is familiar with the farming methods, is in control of the farming activities, and benefits directly from them. On the other hand, local residents are unfamiliar with the farming methods, do not have control over the activities, and experience nuisance without benefits. Risk perceptions of other stakeholders are less pronounced, but are also influenced by factors such as values, cultural background, available knowledge, and potential benefits.

When addressing environmental problems, potential framing differences should be inventoried, made explicit, and – when possible – accounted for. This can avoid potential complications at a later stage. An example from the Angry Bulbs case is the illegal use of pesticides. Local residents are worried about illegal use, but it is very difficult to take this into account in the scientific study. Excluding illegal use implies that not all worries of local residents can be resolved or addressed. This should be a conscious decision that is communicated to all stakeholders beforehand to avoid misinterpretations and disappointments at a later stage. Mapping the problem as perceived by different stakeholders can help to elucidate framing uncertainty. Setting up a dialogue between stakeholders can then reduce framing uncertainty, but this may not always be feasible or necessary. An awareness of framing uncertainty reduces the likelihood of unpleasant surprises.

Knowledge Uncertainty

The Angry Bulb case involves many knowledge uncertainties. The most prominent uncertainty is the question of whether pesticides applied in lily bulb farming can trigger health effects in local residents. This question can be subdivided into questions addressing other uncertainties, such as 'Which pesticides are used by the lily bulb farmers?', 'How high is the pesticide exposure of residents living next to lily bulb fields?', 'What potential health effects can be triggered after pesticide exposure?', 'Are the safety factors applied to the outcome of a rat study sufficiently high to protect humans?', and 'What happens if you are exposed to multiple pesticides (mixtures)?' But there are many more uncertainties involved, e.g. 'Will lily bulb farming in the community of Westerveld continue to grow in the years to come?', and 'How is the application of pesticides regulated?' The latter question illustrates that knowledge is stakeholder-dependent. Most lily bulb farmers and regional authorities probably know how pesticide application is regulated, but local residents typically do not. For them, the threats they experience are a stimulus to gather knowledge on pesticide regulations and health risks. As time proceeds, their knowledge increases until they discover that local residents are not covered in pesticide regulations, revealing a general lack of knowledge applicable to all stakeholders. It is important to note that the process of knowledge-gathering by local residents involves the transfer of knowledge from different sources, e.g. the internet, experts, and scientific literature. Trust in the source plays an important role in this transfer process. This is where knowledge links to risk perception and framing: risk perception is influenced by the knowledge that one trusts, or that comes from a source that one trusts.

Variability

The Angry Bulb case also involves many different types of variability, particularly in the physical domain. Examples include variability in weather conditions or in behaviour, both of which may result in occasional high exposures. Another example is the variability in sensitivity to toxic substances, which can vary with age, genetic background, and health conditions. The models used in risk assessment account for these variabilities by making conservative assumptions, but it is hard to cover each and every extreme situation. The level at which risk is determined is also important in this context. What seems like a small and acceptable risk at a national level – e.g. a one-in-a-million tumour risk – can become highly problematic for you personally if you are living in a highly exposed area and turn out to be very sensitive. What seems acceptable at a population level may thus not always be acceptable at the individual level, particularly if the entity that suffers the adverse consequences has a face.

Besides variability in the physical domain, there is also variability in the social domain. The variation in risk perceptions and problem framings is essentially a form of variability, as is the variability in (access to) knowledge between stakeholders. At a regulatory level, there can be variability in regulations, e.g. between municipalities, regions, or nations. For example, whereas the Dutch assessment procedure for pesticides does not account for local residents, the Danish procedure does. Finally, people involved in the assessment and regulation of pesticides may form a source of variability. For example, it is well known that different experts may interpret the results of scientific studies differently, resulting in different conclusions. This explains why some substances are classified as carcinogenic in the USA, but not in Europe, and vice versa.

Concluding Remarks

The Angry Bulb case nicely demonstrates the role uncertainty can play in environmental issues. The pattern that can be observed in this case is typical for many environmental issues. At its heart lies a conflict of interests between two groups of stakeholders, i.e. local residents and lily bulb farmers. The relationship between lily bulb farming and the threats experienced by local residents is ambiguous and becomes the topic of debate. Uncertainty becomes the playing field of the stakeholders. The stakeholders resort to experts for conclusive evidence and definite answers, which they can rarely provide in complex situations such as the Angry Bulb case. New research may provide (partial) answers but this takes time, and often it is not known how much time. This may be particularly problematic for activities that result in delayed effects, such as the emission of CO_2 and climate change. The precautionary principle was introduced to deal with such situations, but its operationalisation can be problematic because it presupposes a 'threat of serious or irreversible damage' which can be difficult to prove when there is much uncertainty. This is illustrated by the lifting of the ban on metam sodium in the Angry Bulb case. As science often cannot provide immediate and ultimate answers in complex and uncertain environmental issues, a more productive strategy is to create room for deliberation between stakeholders. Acknowledgement of each other's interests and trust are key for such deliberative processes. Although certainty will not be found in a deliberative process, it may open the door for creative solutions, which will benefit all stakeholders. As one of the lily bulb farmers stated in a recent interview: 'We are now attempting to make bulb farming more environmentally friendly. We have to. In this respect, we are grateful to the Angry Bulb Foundation.'

6

Usable Knowledge

WILLEMIJN TUINSTRA, ESTHER TURNHOUT,
AND WILLEM HALFFMAN

6.1 Introduction

Whereas Chapters 4 and 5 discussed knowledge controversies and the limits of scientific knowledge, respectively, this chapter discusses the issue of 'usable knowledge'. Specifically, it looks at the relation of science to decision making: governments, civil servants, and groups of actors deliberating over collective problems, goals, and solutions. A key question that is often asked is how the sciences can best contribute to policy making: with what kinds of attitude or principles, in what kinds of organisations, with what kinds of communication tools? We also see that people get uneasy, either because there is a feeling that scientific knowledge 'is not used', or, on the other hand, that 'science and policy are intertwined too much'.

This chapter will show that the question of what is 'usable knowledge' is more complex than you might expect. For example, one could ask 'usable for whom?' or 'for what?' It is not only the government that makes policy; companies or residence groups also develop plans and make policy. Whenever people come together, set goals, and establish means to achieve these goals within a certain time frame, they are involved in 'policy making' or 'problem solving'. To connect science to such processes of policy making and problem solving, various ideas and strategies have been put into practice. The chapter introduces concepts to characterise and analyse these strategies. It addresses institutional as well as problem-oriented attempts to connect knowledge production and use. What kinds of arrangements lead to 'usable knowledge', or, alternatively phrased, to knowledge that is 'effective' or has 'impact'? And, finally, can we still sensibly figure out some sort of criteria to evaluate the usability and quality of knowledge and of the knowledge production processes?

6.2 What Do Scientific Experts Do?

The activities of scientific experts in environmental decision making vary greatly. According to Mayer et al. (2004), experts can engage in six types of activities for policy: 1) they can *do research and analyse data*; 2) they can *design and recommend* policy actions; 3) they can *offer strategic advice*; 4) they can *mediate*; 5) they can *democratise*; and 6) they can *clarify values and arguments*. Below, we describe these categories, using the example of municipal authorities planning a new neighbourhood (based on Halffman and Broekhans, 2012). Note that these activities all could be performed by policy makers as well.

Research and analyse: experts gather new knowledge. This can involve measurements and experiments, but also surveys and qualitative research. In our planning example, this could be research into soil conditions, environmental impacts, or the housing needs of the envisaged inhabitants.

Design and recommend: experts may also suggest policy alternatives, or suggest alternative courses of action to actors. This could include designs for the neighbourhood, but also recommendations for policy instruments that could be used to bring about sustainability, such as tax cuts for sustainable buildings.

Offer strategic advice: this is what experts do when they suggest overall approaches. In this case, they could suggest overall planning options (a 'development axis' to the next city, or a satellite town) or an approach to planning (e.g. the government as coordinator rather than planning authority).

Mediate: experts often act as go-betweens, facilitators, or coordinators of policy. They can bring together different forms of knowledge, different worldviews, or even interests that have crystallised into opposing camps, and try to integrate or accommodate those camps. In a town-planning process, expert organisations could help develop solutions that allow for multiple use of the same limited space or reconcile conflicting expectations.

Democratise: experts also have a public role to play in making sure knowledge is shared and policy options can be discussed. This is a role expertise can play in the service of public authorities, but also as a challenge to public authorities. In our example, expert organisations could help citizens design alternative plans that can compete with the already developed public plans for the new neighbourhood.

Clarify values and arguments: in this role, expert organisations ask questions about what is at stake for the new neighbourhood, clarify what arguments are being used to defend plans, and perhaps point out inconsistencies or faulty arguments (Flyvbjerg, 1998). This is something of a meta-role, although it can result in concrete policy recommendations. For example, they could question the assumptions underlying the perceived need for a new neighbourhood.

Specific expert advice to policy makers does not have to fall neatly into one of these categories, but may combine elements of several of them. For example, in order to mediate in a planning process stalled in conflict, experts may need to clarify values and arguments, which in turn may require research and analysis of the decision-making process, or of actors' values. In practice, the experts and policy makers involved will try to establish clear and productive working relations. In the context of a project or when serving on a committee, the various people involved will discuss tasks, activities, and responsibilities. They may decide, for example, that the scientific experts will undertake research and supply knowledge, while the others will prepare policy options. Many expert committees or advisory councils operate in this way. Their members cooperate closely with potential knowledge users, they understand what the existing policy frameworks and preferences are, and they know how to work within this context. This enables them to supply usable knowledge and give advice.

However, it is important to note that experts and policy makers working together on environmental issues normally cannot invent their division of tasks from scratch. They enter a setting that has already been shaped by previous experts and past advisory practices, including formal and informal rules and codes of working, as well as a certain understanding of what counts as authoritative knowledge. Users, such as civil servants, managers, or activists, might have implicit expectations with regard to activities of experts. Also, experts have to work within the context of, for example, the established reputation of an advisory organisation or the memory of a mismanaged past crisis (Halffman and Broekhans, 2012). Another issue is that controversies such as those discussed in Chapter 4 make such disagreements among policy makers or within society at large over what to *do* (actions to solve the problem: a *response* from society) change and expand into expert disagreements over what to *expect* (what is the nature and urgency of the *problem itself*? – as assessed by science). This will then influence both the expectations of future activities of experts and the conditions under which experts can provide reliable advice, which may lead to the development of new rules and formats for expert advice (Halffman and Broekhans, 2012).

6.3 Characterising Science–Policy Dynamics

How experts and policy makers meaningfully work together can be examined at different levels, and is also determined by dynamics at different levels. We can look at those interactions at the level of the organisation of a project, such as planning a new neighbourhood or agreeing on a standard for air quality levels in airplane cabins. In those cases, we focus on interactions between individuals. In the planning case, experts and local people interact, while in the cabin case there will be an international group of people who together have to agree on, for example, a European standard. In each case, the tasks and activities of experts can be coordinated, such as by using the classification of Mayer et al. mentioned in the previous section. But one can also look at a much higher level. How is expertise used in governmental policy making in a certain sector? How has the relationship between expertise and policy evolved over the years? For some countries this will vary greatly among policy fields, for other countries there might be a kind of similar line in all policy fields (Halffman, 2005). One can also look at different phases in time. In short, several angles and several lenses might be used to look at the science–policy interface at the same time. In the following we will show how one can look at science–policy interactions depending on the phase of policy making, the characteristics of the problem, and long-term patterns of decision making, respectively.

Firstly, looking at the phase of a policy-making process, we consider whether the issue is still in a problem-defining phase or whether solutions are being developed. As a simplified image of decision making one might consider a cycle that runs from problem identification to policy design, through a political decision, to implementation, and evaluation, back to an assessment of whether the problem has been addressed effectively (Allison and Zelikow, 1971; Drucker, 1967). In the planning case mentioned in the previous section, policy making would start with the observation of a housing shortage, followed by the design of a policy to bring about a new neighbourhood, a political decision to endorse the new neighbourhood, and the actual implementation of this plan by allocating plots or providing permits, then end with an assessment of whether the housing shortage was resolved after the neighbourhood was built (Halffman and Broekhans, 2012). In practice, however, such a neat sequence of phases is rarely the case. For example, disagreement might arise about the definition of the housing shortage once the project is already underway, which might give rise to more, and increasingly complicated, iterations. Still, it is obvious that experts have different work to do in the phase of problem identification than in the phase of assessing whether the problem has been addressed effectively. Depending on the phase, the roles and room for manoeuvre of experts may vary.

Secondly, room for manoeuvre also depends on the way the *policy problem* is perceived or framed: is it considered to be a relatively straightforward problem, or is it a rather ill-defined or so-called 'wicked' problem in which uncertainties and controversies are at the heart of the deliberations? In the case of straight-forward or 'structured' problems there is supposed to be agreement on the relevant knowledge, and people share similar ideas about values and preferred solutions. The solution might be a straightforward technical solution that can be provided by an expert. For example, in the case of noise from a motorway disturbing a neighbourhood, people might agree on building a sound barrier. A problem is 'wicked' when there is disagreement about what the problem is and about what, if any, expert knowledge is required to address the problem (Rittel and Webber, 1973; Hisschemöller and Hoppe, 1996; Hoppe, 2010). Where relevant actors in policy and society disagree not only about the solution but also about the nature of the problem – as is common in current environmental problems such as climate change and biodiversity loss – the role of experts will be much less straightforward, as we will see in Chapter 9. Of course, also in the noise disturbance case the problem might turn out not to be as clear and well-structured as initially conceived. Characteristic of wicked problems is that the way in which they are framed is contested and changes over time.

Thirdly, we can look at how relations between policy and expert organisations have developed over the years. Who dominates the relationship? When experts dominate policy making we speak of a technocratic relationship. This is the case when, for example, engineers, rather than politicians, decide on railway routes. At the other extreme, experts may provide options, but policy makers are in clear control of decisions – including the decision of whether to take any expert advice on board at all. Such long-term and established relations and interaction patterns can exist at the country or sector levels. How is collective decision making organised, and how is expertise organised to accommodate this? What patterns of decision making and advice dominate a policy sector or a country? Here, notable differences in the organisation of expertise have been observed between countries or sectors that are characterised by a consensus or compromise culture and more polarised countries or sectors (Halffman, 2005).

As we have seen, experts and policymakers are typically not completely free to explore and coordinate their roles and task divisions. They will conform to the expectations, commitments, routines, and agreements of the organisations they work for. Some of these agreements may even be anchored in laws and regulations. For example, the European Union has complex procedures in place to establish fish quotas for all EU members. The procedures aim at integrating the scientific advice of fisheries' biologists in negotiations between fisheries' interests and national stakes in fishing grounds. As the fisheries case (Case E: Expertise for European Fisheries Policy) shows, the result may not always be optimal from a conservation point of view, as many biologists and environmentalists insist that fish stocks are still endangered, and hence fishery conflicts escalate. Nevertheless, these procedures and regulations provide at least some sets of rules, tools, and practices to deal with these tensions.

6.4 Strategies to Connect Knowledge Production and Use

One response to the challenge of connecting knowledge production and use is the creation of an intermediary domain in which science, policy, and society are connected and in which usable knowledge is created. In this section, we will discuss two different strategies that are prominent in this intermediary domain: the creation of boundary organisations, and the use of shared concepts or boundary objects.

6.4.1 Boundary Organisations

Boundary organisations are involved in the creation of usable knowledge and play a mediating role between the production of and the use of knowledge. They are accountable to both science and politics, and facilitate the transfer of usable knowledge between science and policy. They also facilitate the organisation of the boundary between the science and policy domain (Guston, 1999, 2001). Typical boundary organisations are advisory councils, specialised government agencies, and consultancies. Examples are the US Environmental Protection Agency (Jasanoff, 1990), the IPCC (Case B: What Does 'Climategate' Tell Us), the IPBES (Chapter 8), and the Dutch Environmental Assessment Agency PBL (Huitema and Turnhout, 2009; Kunseler and Tuinstra, 2017). Situated between knowledge production and knowledge use, boundary organisations are considered capable of navigating the tensions caused by the need for simultaneous proximity and distance between science and policy by means of an internal division of labour. In the ideal case they will be able to identify and meet the needs of policy makers while staying credible with regard to scientific standards. As such, it is assumed that boundary organisations are able to ensure that the expertise produced by these organisations is seen as usable as well as scientifically sound (Pielke, 2007). However, as the fisheries case (Case E: Expertise for European Fisheries Policy) demonstrates, boundary organisations can end up caught in a particular form of thinking or approach that, although it may ensure a smooth collaboration between knowledge producers and users, may not result in the best expertise or policy outcomes.

Boundary organisations not only make connections and bridge the gap between knowledge production and use, they also separate policy makers and knowledge producers. A gap can only be bridged if it is presumed to exist. This implies that boundary organisations need to carefully coordinate their activities with the scientists whose knowledge they synthesise, integrate, and translate, and with the policy makers who make decisions on the basis of the knowledge they provide. As such, rather than internalising the boundary between knowledge production and use, they replace it with two new ones: a boundary between science and the boundary organisation, and a boundary between the boundary organisation and decision making. In this sense, perhaps ironically, boundary organisations play a vital role in keeping science, policy, and society apart; they enable science to strengthen its image as pure and untainted by interests, while still contributing to society through separate organisations and activities. They also enable decision makers to strengthen their image as the ones who are in charge, while at the same time being able to claim that their decisions are based on science and knowledge.

6.4.2 Shared Concepts or Boundary Objects

Another useful concept to help understand how science and decision making are interconnected is the concept of boundary objects. Boundary objects are objects that are developed and used in different domains; they are stable enough to connect different domains, yet at the same time they are flexible and ambiguous enough to allow for different interpretations (Star and Griesemer, 1989). For example, ecological indicators can work as boundary objects. The concept of ecological indication refers to the basic idea that the quality of nature and the environment can be represented using a limited set of measures. It is often up to scientific experts to select those measures. They use their knowledge about the functioning of ecosystems to identify those characteristics of nature and environment – for example, the presence or abundance of certain species or the amount of certain nutrients and pollutants in soils or water systems – that they believe are indicative of specific qualities such as biodiversity, environmental health, or resilience. This also makes clear that ecological indicators are value-laden; they measure the qualities that are considered important (Turnhout et al., 2007). In a policy context, ecological indicators are used to evaluate progress in achieving policy objectives. Consequently, they reflect the values that have informed those policy objectives. The final choice of indicators, for example for nature conservation policy, is the result of interaction and negotiation between scientific and policy considerations. They are stable enough to facilitate interaction between science and policy in the reporting on the state of the art in nature and environment – as assessed by means of the indicators – and the interpretation of this information for policy purposes. At the same time, they will continue to allow for different interpretations by the different users of indicators (Turnhout, 2009). The primary concern for scientists is how the indicators represent nature and environment and how they fit with their research priorities. For policy makers, their main interest is the link between the indicators and the policy objectives and whether the measurements provide a good overview of policy performance.

Another example is the concept of critical loads and how this has been applied in negotiations on European air pollution policy (see Case G: Integrated Assessment for Long-Range Transboundary Air Pollution). A 'critical load' is the maximum exposure to one or several pollutants, at which no harmful effects occur to sensitive ecosystems in the long run (Nilsson and Grennfelt, 1988). Critical loads are used to help to determine effective emission reduction strategies for air pollutants in order to minimise effects on ecosystems. However, even if certain strategies are considered the most effective, in international negotiations it is not always possible to agree on them. Therefore, in one of the negotiation rounds of the United Nations Economic Commission for Europe's Long-Range Transboundary Air Pollution (LRTAP) convention in 1992, the Norwegian delegation proposed a 'gap closure' approach. This concept implies that the 'gap' between current levels of atmospheric deposition of sulphur and the critical loads is 'closed'. The introduction of the concept came at a moment when it was difficult to reach an agreement on the kind of targets for ecological protection to base further negotiations on. The acceptance of the gap closure approach by all parties was a breakthrough in both the science and the policy processes in the preparatory phase of the second sulphur protocol under the convention (1994). The approach was appealing because it formed a direct link to critical loads (which vary with the ecosystems throughout Europe), and because it implied a kind of equity, because the agreed target was a specific percentage of gap closure. The concept offered a way to deal with dissent and formed an important step towards reaching a consensus (Tuinstra et al., 2006).

6.5 What Counts as Usable Knowledge?

In the literature, various sets of criteria for usable knowledge can be found, and these sets partly overlap. Often they combine content criteria with user criteria (e.g. Rich, 1991; Weiss, 1995). Table 6.1 presents an example of such a set of criteria from Weiss (1995).

Studies of knowledge utilisation, including surveys among policy makers, have shown that each of these criteria has been used as a reason why certain knowledge and information has been perceived as useful or not useful in specific circumstances. However, when looking more closely at Table 6.1, we see that it will be difficult to meet all these criteria at the same time. For example, how can knowledge conform to the existing preferences of users

Table 6.1: *Criteria for usable knowledge (Weiss, 1995)*

relevance	knowledge needs to be timely and relate to the topics that are currently salient
conformity	knowledge needs to conform to the prior knowledge, experiences, and beliefs of the users
quality	knowledge should meet scientific standards regarding objectivity, methodology, and accuracy
action-oriented	knowledge should offer a direction for action, for example by presenting alternatives
challenging	knowledge should be interesting and challenging, for example by offering new perspectives or innovative ideas

and be challenging at the same time? There are also other tensions between the criteria. Contact with users cannot be avoided when one wants to develop relevant knowledge. Similarly, conformity requires insight into the beliefs and experiences of those users. However, such close relation-ships with users seem at odds with scientific quality criteria such as objec-tivity (see Chapter 2). Experts focusing fully on 'relevant' knowledge may please civil servants in the short term, but we also want experts to maintain 'independence'. As the case about fisheries policy demonstrates (Case E: Expertise for European Fisheries Policy), such a focus on relevance may result in a situation where civil servants and specialist scientists might get along very well, but where policies are produced that are perceived to be biased and unjust by others because they operate within their own 'bubble', assuming a certain frame of a problem which excludes certain relevant actors and policy options. Next to incompatibilities, there are other reasons why these criteria cannot be applied to evaluate the usability of knowledge in a simple way. The definition of these criteria is often not straightforward, various interpretations might exist, and empirical assessment is difficult. Who decides what counts as high-quality knowledge, action-oriented knowledge, or relevant knowledge?

To conclude, there are two reasons why criteria such as the ones posed by Weiss (1995) cannot be used as 'design principles' for the production of usable knowledge: 1) it is not possible to predict in advance which criteria will be considered important; and 2) it is not possible to determine what knowledge will be seen as relevant, challenging, and so on (Turnhout, 2009). Usable knowledge is no simple matter, and the question of what knowledge is usable crucially depends on the context in which it is developed and applied. This issue will be taken up further in the next section.

6.6 The Importance of Context

As we have seen in Chapter 4, according to the linear model, knowledge is mostly conceptualised as a 'package' that can move from one situation to another. This explains why people often perceive a 'gap' between knowledge that is produced and knowledge that is needed; either the wrong package seems to have been delivered, or the package has not arrived at all. However, such a perspective fails to recognise that knowledge does not move from producers to users without being changed and transformed in the process. This is why knowledge is better considered as a process of interaction characterised by multiple changing meanings and interpretations about what the knowledge is about, and how relevant, challenging, or good it is considered to be (Halffman, 2008). It follows that knowledge produced or used in one specific context is not necessarily usable in other contexts, even if it is, so to speak, 'the same' science. For example, knowledge about acidification of natural ecosystems and systems to collect data in order to monitor and take action to mitigate air pollution is useful in north-west Europe. Acidification is much less of an issue in southern Europe, where drought is more of a problem, as are health problems in cities due to urban air pollution. A similar data-collection and monitoring system as in north-west Europe is less useful here, because different data at a different scale are needed in order to take effective measures to mitigate air pollution. While the knowledge resulting from a pan-European data-collection effort will probably be evaluated as being 'usable knowledge' for north-west Europe, in the southern-European context it will be judged less useful. And, as the fisheries case (Case E: Expertise for European Fisheries Policy) demonstrates, whether knowledge is useful is not just a matter of matching knowledge with nature, but also of matching it with the structure of decision making.

It is therefore not really possible to establish universal criteria for 'usable knowledge'. Instead, as will be detailed in the next section, we can observe a number of ideas, arrangements, and strategies that intend to better connect the production of and use of knowledge.

6.7 Starting from Practice and Looking for Impact

In recent decades, some researchers have sought to approach the production of knowledge from the angle of the problems themselves rather than from a 'science' or 'policy' angle. These approaches include theories, concepts, and methodologies that are often accompanied by guidelines, design principles, and rules of thumb. Examples are action research, transdisciplinary research, 'integration and implementation science', and, more recently, the further development and operationalisation of concepts such as 'research impact' and 'knowledge exchange'. These various approaches have their origin in different disciplines and policy fields, but have in common that they stress the importance of mutual learning among the various societal groups involved in the problem. Knowledge development is targeted at solving the problem at hand and explicitly involves integration of various kinds of knowledge and scientific disciplines.

Action research, for example, actually originated in the 1950s in the field of psychology, and developed further in the 1970s, mainly in the field of agriculture and development. It is targeted at social change and involves local groups and external experts collaborating to find optimal solutions, such as for improving the sustainability of local water and sanitation services (Wadsworth, 1998; Burns, 2012).

Transdisciplinary research has been defined as 'a new form of learning and problem solving involving co-operation between different parts of society and science in order to meet complex challenges of society. Transdisciplinary research starts from tangible real-world problems. Solutions are devised in collaboration with stakeholders' (Klein et al., 2001, 7). Or, as formulated by the Swiss Academies of Arts and Science: 'Transdisciplinary research deals with problem fields in such a way that it can a) grasp the complexity of problems, b) take into account the diversity of perceptions, c) link abstract and case-specific knowledge, and d) develop knowledge and practices that promote what is perceived to be the common good' (Pohl and Hirsch Hadorn, 2007, 20).

For example, in a transdisciplinary research project in the Swiss canton Appenzell Ausserboden, inhabitants, civil servants, and scientists and students from various research institutes have collaboratively been analysing different industry sectors (timber, dairy farming, and textile production) and different aspects of regional development (land use, mobility, landscape protection, and tourism) in order to generate cross-sectoral long-term development strategies dealing with the cantons' problems of structural change and migration (Walter et al., 2008). The Swiss Academies of Arts and Science proposes four design principles to take into account when shaping the transdisciplinary research process: 1) reduce complexity by specifying the need for knowledge and identifying those involved; 2) achieve effectiveness through contextualisation; 3) achieve integration through open encounters (collaboration through exchange of perspectives); and 4) develop reflexivity through recursivity (iterate and adapt) (Pohl and Hirsch Hadorn, 2007, 20–23). Next to problem identification and problem analysis, transdisciplinary research stresses the importance of 'bringing results to fruition'.

Integration and Implementation Science has a similar ambition of applicability: tackling social and environmental problems by enhancing 1) synthesis of disciplinary and stakeholder knowledge; 2) understanding and management of diverse unknowns; and 3) provision of integrated research support for policy and practice change (Bammer, 2013). In addition, it searches for ways to mainstream this kind of research as a 'discipline' in academia, because interdisciplinary researchers still feel hampered by e.g. inadequate assessment mechanisms for tenure and promotion and lower levels of success in grant applications (Bammer, 2017).

Reasoning from a researchers' perspective as well, while equally concerned with the 'impact' or 'good that researchers can do in the world', is the community of scholars focusing on the operationalisation of the concept of 'research impact' and its precursor 'knowledge exchange' (Reed, 2018; Fazey et al., 2013). Knowledge exchange processes are processes that generate, share, and/or use knowledge through various methods appropriate to the context, purpose, and participants involved. Knowledge exchange includes concepts such as sharing, generation, co-production, co-management, and brokerage of knowledge and is closely related to learning (Fazey et al., 2013). Reed (2018) notes five important factors in the impact of research: 1) the context and purpose of the research; 2) who initiated the research; 3) the inclusion of relevant perspectives and interest; 4) the way the design of the research takes these first three factors in account; and 5) the way the design takes power dynamics into consideration. Important design principles again include problem and solution orientation, the integration of various sources of knowledge, and participation. In Chapters 7 and 8 we will explore these issues further. We will see that integration of knowledge comes with its own challenges, and also that participatory forms of knowledge production introduce some crucial problems of their own.

6.8 Credibility, Salience, and Legitimacy

Finally, we have to revisit the issue of the evaluation of usable knowledge. In this chapter, we have seen that the activities of scientific experts greatly vary across various processes, that they interact with decision makers on various levels, and that the boundaries drawn between science and decision making also may vary. Also, what is usable for people within a process might not seem useful for others who are not included in that process. Moreover, it is clear that the usability of knowledge is not an attribute of 'the knowledge' in and of itself, but is determined in context and shaped by the processes of interaction involved in the production and use of knowledge (Turnhout, 2009). From this perspective, we have to replace criteria to evaluate knowledge – such as the ones we discussed in Section 6.5 – with criteria we can use to evaluate these processes.

One way to do this is to evaluate advisory processes or knowledge production processes in terms of credibility, salience (relevance), and legitimacy (Farrell et al., 2001; Cash et al., 2003; Farrell and Jäger, 2005). Credibility refers to perceptions of the degree to which scientific quality criteria are met, whether valid methods are used, and whether results and conclusions have a firm and authoritative foundation. Salience refers to relevance for users, including the link to agendas of decision makers, the accessibility of knowledge, the right timing, and a clear connection with the choices available to actors. Legitimacy has the most outspoken process component. It refers to the perceived fairness of the knowledge production process, to who is involved, and to the extent to which the concerns and perspectives of various actors have been considered (Farrel et al., 2001). Attention to the legitimacy dimension of knowledge production can help overcome problems of inclusion and exclusion and can prevent certain forms of expertise coming to dominate. However, no straightforward way exists to ensure credibility, salience, and legitimacy. Not only will different actors evaluate and interpret credibility, salience, and legitimacy in different ways, there are also potential trade-offs between the three criteria. For example, a strong focus on credibility can pose a threat to legitimacy if credibility is interpreted in a relatively strict scientific way that excludes other forms of knowledge and limits participation and engagement. Thus, usable knowledge-making concerns the collective deliberation and negotiation of acceptable levels of credibility, salience, and legitimacy. As such, they are emerging properties of the *process* of the production of usable knowledge, not attributes of knowledge itself.

References

Allison, G. T., and Zelikow, P. (1971). *Essence of Decision*. Boston: Little, Brown and Company.
Bammer, G. (2013). *Disciplining Interdisciplinarity: Integration and Implementation Sciences for Researching Complex Real-World Problems*. Canberra: ANU E-Press. doi: 10.22459/DI.01.2013
Bammer, G. (2017). Should We Discipline Interdisicplinarity? *Palgrave Communications*, 3 (30). doi: 10.1057/s41599-017-0039-7
Burns, D. (2012) Participatory Systemic Inquiry. *IDS Bulletin*, 43(3), 88–100.
Cash, D. W., Clark, W. C., Alcock, F., et al. (2003). Knowledge Systems for Sustainable Development. *Proceedings of the National Academy of Sciences of the United States of America* 100(14), 8086–8091.

Drucker, P. F. (1967). *The Effective Decision. Harvard University, Graduate School of Business Administration.* Cambridge, MA: Lansford.

Farrell, A. E., VanDeveer, S. D., and Jäger, J. (2001). Environmental Assessments: Four Under-appreciated Elements of Design. *Global Environmental Change* 11(4), 311–333.

Farrell, A. E., and Jäger, J., eds. (2005). *Assessments of Regional and Global Environmental Risks: Designing Processes for the Effective Use of Science in Decisionmaking.* Washington, DC: RFF Press.

Fazey, I., Evely, A. C., Reed, M. S., et al. (2013). Knowledge Exchange: A Review and Research Agenda for Environmental Management. *Environmental Conservation* 40, 19–36.

Flyvbjerg, B. (1998). *Rationality and Power: Democracy and Practice.* Chicago: University of Chicago Press.

Guston, D. H. (1999). Stabilizing the Boundary Between US Politics and Science: The Role of the Office of Technology Transfer as a Boundary Organization. *Social Studies of Science*, 29, 87–111.

Guston, D. H. (2001). Boundary Organisations in Environmental Policy and Science: An Introduction. *Science, Technology and Human Values*, 26, 399–408.

Halffman, W. (2005). Science/Policy Boundaries: National Styles? *Science and Public Policy*, 32(6), 457–467. doi: 10.3152/147154305781779281

Halffman, W. (2008). *States of Nature: Nature and Fish Stock Reports for Policy.* The Hague: Netherlands Consultative Committee of Sector Councils for Research and Development.

Halffman, W., and Broekhans, B. (2012). The Landscape of Environmental Expertise. In W. Tuinstra, ed., *Course Book Environmental Problems: Crossing Boundaries Between Science, Policy and Society* (pp. 55–87). Heerlen: Open Universiteit.

Hisschemöller, M., and Hoppe, R. (1996). Coping with Intractable Controversies: The Case for Problem Structuring in Policy Design and Analysis. *Knowledge and Policy*, 8, 40–60.

Hoppe, R. A. (2010). *The Governance of Problems: Puzzling, Powering, and Participation.* Portland: The Policy Press.

Huitema, D., and Turnhout, E. (2009). Working at the Science–Policy Interface: A Discursive Analysis of Boundary Work at the Netherlands Environmental Assessment Agency. *Environmental Politics*, 18, 576–594.

Klein, J. T., Grossenbacher-Mansuy, W., Haberli, R., Bill, A., Scholz, R. W., and Welti, M. (2001). *Transdisciplinarity: Joint Problem Solving among Science, Technology, and Society.* Basel: Birkhauser Verlag.

Jasanoff, S. (1990). *The Fifth Branch: Science Advisers as Policymakers.* Boston: Harvard University Press.

Kunseler, E.-M., and Tuinstra, W. (2017). Navigating the Authority Paradox: Practising Objectivity in Environmental Expertise. *Environmental Science and Policy* 67, 1–7. http://dx.doi.org/10.1016/j.envsci.2016.10.001

Mayer, I. S., Daalen, C. E. van, and Bots, P. W. G. (2004). Perspectives on Policy Analyses: A Framework for Understanding and Design. *International Journal of Technology, Policy and Management*, 4(2), 169–191.

Nilsson, J., and Grennfelt, P., eds. (1988). *Critical Loads for Sulphur and Nitrogen.* The Nordic Council of Ministers 15. Copenhagen.

Pielke, R. (2007). *The Honest Broker Making Sense of Science in Policy and Politics*. Cambridge: Cambridge University Press.

Pohl, C., and Hirsch Hadorn, G. (2007). *Principles for Designing Transdisciplinary Research Proposed by the Swiss Academies of Arts and Sciences*. München: Oekom Verlag.

Reed, M. S. (2018). *The Research Impact Handbook*, 2nd edn. Newcastle upon Tyne: Fast Track Impact.

Rich, R. F. (1991). Knowledge Creation, Diffusion and Utilization, Perspectives of the Founding Editor of Knowledge. *Knowledge Creation, Diffusion and Utilization*, 12, 319–337.

Rittel, H., and M. Webber (1973). Dilemmas in a General Theory of Planning. *Policy Sciences*, 4, 155–169.

Star, S. L., and Griesemer, J. (1989). Institutional Ecology, 'Translations', and Boundary Objects: Amateurs and Professionals in Berkeley's Museum of Vertebrate Zoology, 1907–1939. *Social Studies of Science*, 19, 387–420.

Tuinstra, W., Hordijk, L., and Kroeze, C. (2006). Moving Boundaries in Transboundary Air Pollution Co-production of Science and Policy under the Convention on Long Range Transboundary Air Pollution. *Global Environmental Change*, 16, 349–363.

Turnhout, E. (2009). The Effectiveness of Boundary Objects: The Case of Ecological Indicators. *Science and Public Policy*, 36, 403–412.

Turnhout, E., Hisschemöller, M., and Eijsackers, H. (2007). Ecological Indicators: Between the Two Fires of Science and Policy. *Ecological Indicators*, 7(2), 215–228.

Wadsworth, Y. (1998). What Is Participatory Action Research? Action Research International Paper 2. www.aral.com.au/ari/p-ywadsworth98.html

Walter, A.I., Wiek, A., and Scholz, R.W. (2008). Constructing Regional Development Srategies: A Case Study Approach for Integrated Planning and Synthesis. In Hirsch Hadorn, et al., (eds.) *Handbook of Transdisciplinary Research* (pp. 223–243). Dordrecht: Springer.

Weiss, C. H. (1995). The Haphazard Connection: Social Science and Public Policy. *International Journal of Educational Research*, 23, 137–150.

Case E Expertise for European Fisheries Policy
WILLEM HALFFMAN AND MARTIN PASTOORS

The fisheries case study shows how relevant, timely, and actionable expertise develops over a period of decades in a concrete policy field. It shows how such arrangements get embedded in organisational procedures and even mathematical tools, the so-called Total Allowable Catch Machine (see also Chapter 7). The case illustrates how the usefulness of expert advice depends on for whom and for what the advice is supposed to be useful. In fisheries policy, this usefulness was historically dominated by concerns for fishing-right conflicts, and far less by concerns for sustainability. Some science advice has been 'useful' in criticising the assumptions of this system.

Fisheries in Perspective

Fisheries policy may sound like an exceptional or even unimportant policy field, but such first impressions would be deceiving. In Europe, the fishing industry is a sector of considerable importance, both economically and politically. With a European annual turnover of a 'mere' €28 billion, it is nevertheless a crucial source of income and employment in many coastal regions (European Commission, 2016). In addition, disgruntled fishers know how to appeal to national sentiments and how to make themselves heard. For example, with the promise of reclaiming exclusive access to UK waters, British pro-Brexit fishing boats campaigning on the London Thames received extensive media coverage in the run-up to the referendum (Payton, 2016). French and Spanish fishers also have a militant spirit and have organised extremely expensive harbour blockades in the past to put pressure on their governments and, subsequently, on European fishing authorities (Mackenzie, 2009). While relatively small, the fishing industry does therefore have considerable leverage.

Fisheries policy is also of wider importance for our understanding of environmental policies in general. Fish stocks provide a classic and much-discussed example of the tragedy of the commons, the particular political/economic logic behind environmental degradation originally articulated by Hardin (1968): a shared natural resource that is finite, and that gets appropriated when collected, tends to be depleted if harvesting possibilities are allowed to outgrow resource recovery rates. Or, as illustrated for fish: without some form of limitation, our fishing fleets and their technologies will grow so fast, competitively trying to catch fish before anyone else does, that fish stocks will be at risk of fatal depletion.[1]

The history of fisheries offers various examples of more or less dramatic fish stock depletions, some going back as far as the Middle Ages. In some cases, collapsed fish stocks even lead to altered ecological equilibriums, to which the depleted species may not return for decades. A notorious example is the collapse of the Newfoundland cod in the early 1990s, in which governments, warned by fishers and fisheries biologists, hesitated just a little too long to restrict catches. By the time the Canadian government implemented a complete moratorium on cod fishing in 1992, the cod was gone. After hundreds of years of fishing on the Grand Banks, the cod fishing industry folded, taking 30,000 livelihoods with it and devastating local fishing communities. A decimated fleet of remaining fishers had to shift to other species, such as crab. A small amount of cod could be caught by the end of the 1990s, partly to get a sense of the stock's condition, but only very recently, after a quarter-century, have fisheries regulators started to consider the possibility that cod stocks might recover (Davis, 2014; Kurlansky, 1997).

The fact that some fish stocks have been managed successfully, across the globe and throughout history, has also clarified the conditions under which sustainable management of fish stocks is possible – although there is no simple solution. The description and articulation of such conditions, controlling access to fisheries and setting up restrictions that take into account power imbalances such as those between small and larger fishers, has led to major improvements in our understanding of sustainable economics – and even a Nobel Prize (Ostrom, 1990). In addition to their economic and political importance, fisheries are also a textbook case for sustainability policy.

Any fisheries governance system has to deal with the problem of how to assess the state of the environment, and a key complication is that assessing fish stocks involves major uncertainties. It is notoriously hard to establish the size of a fish population at sea, simply because the ocean is huge, fish populations are unequally distributed, and sampling requires expensive ships and equipment. A further complication is that sustainable fish harvesting not only needs

---

[1] The tragedy of the commons is illustrated in the fisheries simulation game *Fishbanks*, originally developed by Denis Meadows, and an excellent way to demonstrate the process (Meadows et al., 2017).

to establish present fish populations, but also *future* fish populations: in order to establish fish quota, fishery biologists need to predict fish stocks a year in advance, and preferably longer. This could imply assessing population size, but also population age structure (which may affect reproduction rates), ideally also predator/prey relations, or other changes in the marine environment. Meanwhile, fish populations are affected by the most difficult of all factors to predict: human behaviour, including expected market development, fishing success, expected investments, effective adherence to quota, and changes in policy (Halffman, 2008).

Such predictions are bewilderingly complex and forever fraught with uncertainties, even in retrospect. In the case of Newfoundland, there is continued debate over the relative contribution of various factors in the cod collapse: overfishing by Canadian and foreign fleets, a decline in capelin (the typical prey species for cod), predation by a grown seal population, sea temperature changes, and disturbances by oil and gas exploitation (Davis, 2014). Fisheries biologists largely have to rely on catch data to assess stocks, but fishers constantly improve technologies, confounding an indication of the population with an indication of the fishers' prowess to catch fish. Another major complication is that surveillance of fishing and restrictions on equipment are very hard to implement out at sea, and what fishers land and report at their home ports is not always what was actually caught. 'Illegal, unreported and unregulated fishing', such as by transfers at sea, is estimated at more than 15 per cent of global catches (European Commission, 2016), and much higher by environmental campaigners. Meanwhile, model predictions are presented as if they were highly precise, without adequate expression of the uncertainties involved. The numbers produced by complex fishing models to predict stocks may therefore present policy makers with a misleading sense of certainty (Halffman, 2008; Hauge, 2011; International Council for the Exploration of the Sea, 2006).

From this complex problem, with many interacting processes and uncertainties that are hard to pin down and disentangle, fisheries scientists have to distil clear information for policy makers and knowledge that is relevant and usable, and must do so in time for specific moments such as periodic negotiations over fishing quota. This knowledge has to be 'packaged' and presented in just the right way to be useful in the negotiations. What these negotiations look like, in turn, depends on the fisheries management approach used in policy. If policy relies on fishing quota for individual species set on a yearly basis, then the advice has to come in the form of a recommended catch tonnage per species, specified for the period fishing rights are allocated. If policy is determined via multi-year plans, restrictions in the mesh size of nets, price schemes, or the establishment of no-fishing zones, then the scientists need to provide, respectively, long-term predictions, technological advice, recommended prices, or zoning information. In other words: out of the complex, multi-factor puzzle of what will most likely happen to fish out at sea, the experts have to establish an account that corresponds to how fisheries management is done. Obviously, the experts may also present information that does *not* fit the dominant management scheme, for example to assess whether it is effective or even question its assumptions, but at least the routine management of fisheries requires a correspondence between the expertise and the management approach.

Origins and Operation of EU Fisheries Policy

Historically, European fisheries policy is rooted in the pacification of international conflicts over resources, a concern that also motivated its precursor organisations. (For example, the European Coal and Steel Community was formed after the Second World War to help prevent European wars over mineral resources.) That such concerns were not hypothetical became clear during the so-called Cod Wars, a series of diplomatic and even naval skirmishes over fishing rights in the North Atlantic, particularly between Iceland and the UK. Countries had gradually expanded their territorial waters from 4 nautical miles to the 200-mile zone claimed in 1975, partly motivated by claims over marine oil and gas resources, but also in an attempt to protect threatened fisheries which were vital to the Icelandic economy. With this expansion, historic fishing rights and agreements, sometimes going back to the fifteenth century, were withdrawn and countries expulsed foreign vessels from their newly acquired waters. In the case of Iceland, this included a huge swathe of rich North Atlantic fishing grounds, which British fishermen refused to give up without a fight. On several occasions, the Icelandic coastguard patrol vessels attempted to enforce Iceland's exclusive fishing rights by chasing out British trawlers, and eventually also by attempting to run over and destroy nets. Nationalist sentiments ran high, and British frigates were deployed to escort and protect British fishing in Icelandic waters. A series of cat-and-mouse skirmishes ensued with the far weaker Icelandic patrol vessels, involving obstruction, mutual ramming, and eventually even warning shots. Miraculously, only one sailor lost his life, but several ships were severely damaged, and Iceland cut diplomatic ties with the UK. NATO had to step in and negotiate a deal, as its Cold War strategic positions came under threat (Steinsson, 2016).

While the UK–Iceland Cod Wars are a more extreme example of the potential conflict hiding in the murky Atlantic, tensions between European nations over fish are not uncommon. The European Communities' (EC) fisheries policy offered a forum for fish quota allocation and to organise mutual access to fishing grounds, preventing such escalations. The UK had accessed the EC in 1973, and the renewed 1977 European fisheries policy started a process of delicate negotiations over fishing rights, flanked by the sweetener of fleet modernisation subsidies. It is important to bring this background to mind, because it shaped the backbone of the European fisheries policy problem: how to allocate access to fish stocks between nations while preventing conflict.

Because of the structure of European marine ecology, the specialised organisation of the fishing industry built on it, and the fisheries research that informed it, these negotiations were organised around individual species quota, determined per section of European waters (Figure E1). By 1983, and after long and difficult negotiations, European fisheries policy was organised around the biannual establishment of fishing quota in the form of a Total Allowable Catch (TAC) per 'stock', a regional population of an individual species. The advised TAC was the input for negotiations between national fisheries ministers that would set actual quota per species, per sector, and per country. This meant that twice a year, the negotiation between fisheries ministers of member countries, representing their fishing industry, had to be supplied with expert information that analysed fisheries in this particular way: it *framed* fisheries as a set of individual species populations, with half-year predictions and recommendations based on an assessment of the stock. Because of economic, ecological, but also geopolitical reasons, the complex problem of how to manage fisheries was structured in a very specific way: out of all the interactions in marine ecosystems, economies, and politics, the TAC process defined what counted as usable knowledge (Daw and Gray, 2005; Halffman, 2008).

However, this is not just a story of politics determining the utility of knowledge. Fisheries science too provided part of the template that structured the problem of EU fisheries. The knowledge needed to set TACs is gathered by the International Council for the Exploration of the Sea (ICES), based in Copenhagen. This organisation was established in 1902, as part of an early effort to provide knowledge that could inform policy and the fisheries industry and provide knowledge that could avoid overfishing. ICES is essentially a network organisation, bundling the resources of fisheries research institutes of its member countries. Based on insights in fish population dynamics, essentially revolving around the reproductive success of different age groups in fish populations, they developed an understanding of how a stock would develop over time. The arrival of new modelling techniques allowed them to estimate future population sizes for commercially interesting species, and by 1965 a mathematical tool was developed to derive a TAC from such a model

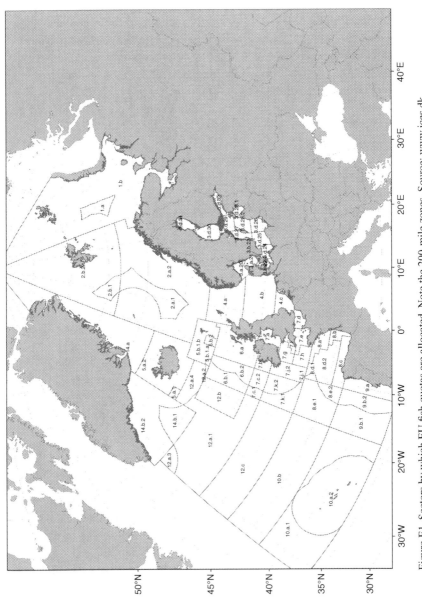

Figure E1 Sectors by which EU fish quotas are allocated. Note the 200-mile zones. Source: www.ices.dk.

(Rozwadowski, 2002, 2004). In the context of the fisheries in European waters, this made more sense than it would have for tropical environments, with a much higher variety of fish species (Kaiser et al., 2005). With the modelling techniques available at the time, a single-species population model was already quite an achievement. The price to pay for making a doable model was that some of the complicating factors of a marine system had to remain largely out of focus, such as predator/prey interactions.

Limitations, Criticism, and the Future of European Fisheries

By the 1990s, this production of usable fish stock knowledge was in full swing, but cracks were beginning to show. In spite of attempts to limit the catch, some stocks still showed signs of overfishing, and sometimes painful emergency measures were needed. Emergency measures meant drastic fluctuations in fishing fleet returns, endangering the long-term financial security of companies having to pay off loans and fixed costs. ICES working groups that brought together scientists from all over Europe to establish scientific agreement over recommendations failed to generate trust in the industry. Under pressure from national fishing industries, negotiating minsters would water down the advised TACs to quota that kept fishers happy. In turn, there were rumours that the fisheries scientists would lower their recommendations in anticipation of political negotiations resulting in higher quota, which then further undermined the credibility of advised TACs. All this time, predictions suffered from the limitation of single-species population dynamics and uncertainty in the data. While the TAC system managed to avoid escalation of international conflicts over fish resources, some European fish stocks veered dangerously towards overfishing thresholds. High fishing pressure kept fish populations small. ICES scientists argued in favour of quota reductions to let populations grow, so that larger returns could be created in the future. A 'maximum sustainable yield' could therefore not only protect fisheries from collapse, but actually maintain larger fish populations, thereby guaranteeing more sustained and durable fisheries, but the decision-making system failed to generate the required trust and cooperation between stakeholders and their experts to really make a difference (Holm and Nielsen, 2004; Kooiman et al., 2005).

This situation led to severe criticism from several sides. It was clear that the TAC system did not lead to sustainable fisheries. Environmentalists started to campaign for more sustainable fisheries and suggested radical alternatives to

TAC-based management, such as marine reserves or sustainability labels alerting consumers. Some fisheries scientists pointed out that social and economic processes were not sufficiently addressed in setting fish quota and that the social sciences that could have elucidated such processes were insufficiently represented in fisheries expertise. They argued that fisheries should be managed via more interaction with stakeholders, take into account fishers' need to plan investment longer than a year ahead, and develop more robust approaches in the face of socio-ecological complexities (Halffman, 2008).

However, in spite of obvious shortcomings, the regulatory approach seemed stuck in a rut: it kept on churning out biannual TAC quota, implemented half-heartedly by policy makers, without achieving longer-term sustainability. Because European fisheries institutions seemed unable to break out of this routine, this was dubbed the 'TAC machine' (Holm and Nielsen, 2004). EU fisheries policy ended up in the literature as an example of policy failure (Daw and Gray, 2005). Grown out of specific concerns and historic conditions, the particular approaches had become embedded in fisheries models, the structure of ICES and its working groups, the decision-making committees, and the routines and regulations, as well as a game of bargaining and anticipation that actors seemed unable to change.

Part of the problem was that the decision-making procedure organised expertise in a linear model (see Figure E2 and Chapter 4 on controversies). ICES was positioned to produce expert assessments based on the science of its population models. The advice was delivered ready-made in stock assessment reports that were supposed to pin down what counted as hard facts for the bargaining game between national ministers that followed. This particular way of organising the process prevented a deliberation over its assumptions (cf. Halffman and Hoppe, 2005). In fact, as it was designed to stabilise negotiations and to prevent escalation of fisheries conflicts, the strategy was to depoliticise some of the conflict by letting experts establish the facts. For its particular goal of pacification, the policy was actually relatively successful: occasionally, harbours were blocked, but nobody had sent out the navy to ram ships in a while.

Does the TAC Machine 'Work'?

ICES can be considered as a boundary organisation that connects fisheries science to decision making (see Chapter 5). The stock assessment models of ICES aim to facilitate the connection between fisheries research and policy. By defining what constitutes usable knowledge, these models function as boundary

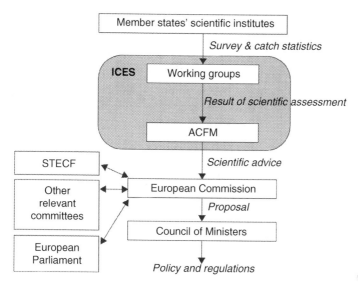

Figure E2 Decision-making structure of the 'TAC machine'. Note the linear structure of the advisory process (STECF: Scientific and Technical Committee on Fisheries, the Commission's own expert committee; ACFM: Advisory Council on Fisheries Management, with stakeholders) (Daw and Gray, 2005).

objects that integrate scientific knowledge to become relevant for decision making. While this arrangement ensures the effective production of, and use of, knowledge, it can only do so because of the way in which the fisheries problem is framed: as a set of individual species populations, with half-year predictions and recommendations based on sector stock assessments and by excluding knowledge and perspectives that do not fit in that frame.

The story of fisheries shows the difficulty of developing institutions that provide usable knowledge. On the one hand, the TAC machine seemed to attune experts and policy makers quite adeptly. Developed over a period of decades, it delivered knowledge right on time, in the form of individual species stock assessments, matching the policy approach that relied on national quota allocation. In that sense, the machine 'worked'. On the other hand, the usefulness of the knowledge was also limited as it did not engender trust between policy makers and experts, did not lead to long-term sustainability, and did not allow for second-order learning that could develop different approaches and think outside of the box (for more detail, see Halffman, 2008). Thus, although the current arrangement could be said to score relatively well on salience, its credibility and legitimacy are being questioned by critics of the current system, including several dissident fisheries scientists.

Over the last decade, important modifications have been made to European fisheries policy. Regionalisation has split up the formerly centralised European system with regional management committees, making it easier to interact with stakeholders. Quota allocation now attempts to have a perspective of five years, dampening quick changes to ease business planning. The management of some stocks takes more complex ecological interactions into account. Marine reserves and other alternative policy instruments are being tried out, although with limited application in European waters. While fish quota remain a key instrument and several stocks remain over-exploited (European Commission, 2016), perhaps the TAC machine has begun to break out of its routine. However, the international conflicts over access to fisheries it was intended to pacify have not disappeared. If the Brexiteers follow through with their promise to take British waters out of the Common Fisheries Policy, a new round of conflicts could erupt.

References

Davis, R. (2014). A Cod Forsaken Place?: Fishing in an Altered State in Newfoundland. *Anthropological Quarterly*, 87(3), 695–726.

Daw, T., and Gray, T. (2005). Fisheries Science and Sustainability in International Policy: A Study of Failure in the European Union's Common Fisheries Policy. *Marine Policy*, 29, 189–197.

European Commission. (2016). *Facts and Figures on the Common Fisheries Policy: Basic Statistical Data, 2016 Edition.* https://ec.europa.eu/fisheries/sites/fisheries/files/docs/body/pcp_en.pdf

Halffman, W. (2008). *States of Nature: Nature and Fish Stock Reports for Policy.* The Hague: RMNO.

Halffman, W., and Hoppe, R. (2005). Science/Policy Boundaries: A Changing Division of Labour in Dutch Expert Policy Advice. In S. Maasen and P. Weingart, eds., *Democratization of Expertise? Exploring Novel Forms of Scientific Advice in Political Decision-making* (pp. 135–152). Dordrecht: Kluwer.

Hardin, G. (1968). The Tragedy of the Commons. *Science*, 162, 1243–1248.

Hauge, K. H. (2011). Uncertainty and Hyper-precision in Fisheries Science and Policy. *Futures*, 43(2), 173–181.

Holm, P., and Nielsen, K. N. (2004). *The TAC Machine, Appendix B, Working Document no. 1 to the 2004 ICES Annual Report of the Working Group for Fisheries Systems.* www.ices.dk.

International Council for the Exploration of the Sea. (2006). *Fish Stocks: Counting the Uncountable?* www.ices.dk.

Kaiser, M. J., Attrill, M. J., Jennings, S., et al. (2005). *Marine Ecology: Processes, Systems, and Impacts.* Oxford: Oxford University Press.

Kooiman, J., Bavinck, M., Jentoff, S., and Pullin, R., eds. (2005). *Fish for Life: Interactive Governance for Fisheries*. Amsterdam: Amsterdam University Press.

Kurlansky, M. (1997). *Cod: A Biography of the Fish that Changed the World*. New York: Penguin Books.

Mackenzie, J. (2009). French Fishermen Maintain Channel Port Blockade. http://uk.reuters.com/article/uk-france-ports-idUKTRE53E58P20090415

Meadows, D., Sterman, J., and King, A. (2017). Fishbanks: A Renewable Resource Management Simulation. https://mitsloan.mit.edu/LearningEdge/simulations/fishbanks/Pages/fish-banks.aspx

Ostrom, E. (1990). *Governing the Commons*. New York: Cambridge University Press.

Payton, M. (2016, 30 June). UK Fishermen Warned Catch Quotas May Not Increase with Brexit. *The Independent*. www.independent.co.uk/news/uk/politics/uk-fisherman-warned-catch-quotas-may-not-increase-with-brexit-a7110766.html

Rozwadowski, H. M. (2002). *The Sea Knows no Boundaries: A Century of Marine Science under ICES*. Seattle: University of Washington Press.

Rozwadowski, H. M. (2004). Internationalism, Environmental Necessity, and National Interest: Marine Science and Other Sciences. *Minerva*, 42(2), 127–149. doi:10.1023/b:mine.0000030023.04586.45

Steinsson, S. (2016). The Cod Wars: A Re-analysis. *European Security*, 25(2), 256–275. doi:10.1080/09662839.2016.1160376

7

Interdisciplinarity and the Challenge of Knowledge Integration

ESTHER TURNHOUT

7.1 Why Integrate?

The domain of environmental policy and governance occupies a rather special position in public policy making processes due to its historical reliance on science as the main source of information to signal environmental problems, assess their character and causes, and suggest solutions. In other words, we could say that decision making for environmental issues has a strong technocratic tradition, in which decision makers look to science for answers and solutions (this has been discussed as the linear model in Chapter 4). To effectively address the multidimensional character of environmental issues, there is a strong call to provide an integrated and interdisciplinary perspective that combines not just natural science, but also social science knowledge. In this chapter, we will introduce a number of common methods or approaches to knowledge integration and critically reflect on their strengths as well as their limitations. But before we do this, we need to delve a bit deeper into the question of why knowledge integration is considered important.

A first common argument for knowledge integration is the complexity of environmental problems. Environmental problems are characterised by complex interlinkages between biotic and abiotic soil, water, and atmospheric systems. This suggests that environmental problems can only be understood if different scientific disciplines such as ecology, soil science, hydrology, meteorology, and so on work together. Second, environmental problems also have an important social dimension. Industry, transport, agriculture, and other human activities affect the environment and are often seen as important causes of environmental problems. This makes clear that social science disciplines also have to be included. Social science knowledge about industrial growth, transportation, or food production is as crucial to understanding environmental problems as natural science knowledge. Thus, in this perspective, integrating knowledge from different disciplines is considered important to create a comprehensive picture that covers all relevant aspects of the problem.

The second key impetus for knowledge integration is related to how to solve environmental problems. To trigger appropriate responses, policy makers need to be presented with an integrated representation of the problem that can be translated into effective policies and measures. To make this link, interdisciplinary cooperation is again important to provide an integrated assessment of the potential effectiveness of different policy options in solving environmental problems.

When it comes to the integration of scientific and non-scientific knowledge (a topic that will be discussed in more detail in Chapter 8), there is an important third argument for knowledge integration that has to do with participation and inclusion. Different actors within and outside science have stakes in the issues and problems that integrated knowledge is supposed to address. These actors also may have relevant knowledge that they want to bring to bear upon these issues and problems, and therefore when this knowledge is used, the solutions that are derived from it will ostensibly enjoy support and buy-in. Thus, the integration of all forms of knowledge that are considered relevant is important not just for the quality of knowledge, but also for the legitimacy of solutions and the effectiveness of their implementation.

These different rationales of knowledge integration – creating the full picture and ensuring that knowledge offers relevant and effective solutions – often occur in conjunction. For example, Mitchell et al. (2006, 3) explain that environmental assessments integrate knowledge about 'scientific and social processes and … how these processes relate', and make this knowledge 'available in a form intended to be useful for decision making'. In a similar vein, Bammer (2005, 1) argues for the development of a new specialised field of science called 'integration and implementation sciences': 'new research skills must be developed if human societies are to be more effective in tackling the complex … problems that confront us … Researchers must collaborate and integrate across traditional [disciplinary] boundaries.'

These integrated forms of knowledge have significant impact because they shape what knowledge is presented to policy processes and therefore also what aspects of the environment can be considered for protection (Miller, 2007). As we will discuss in more detail later in this chapter, knowledge integration comes at a price as different forms of knowledge are not easily combined and this means that interdisciplinary collaboration is quite difficult. In view of the potential political impact of integrated knowledge and the challenges involved in its production, it is important to critically reflect on what knowledge integration means and how it is done in practice.

7.2 What Is Interdisciplinarity?

Before we move to discussing how knowledge from different disciplines can be integrated in practice, we need to reflect for a moment on the idea of discipline. As discussed in Chapter 2, disciplines are one of the ways in which science is organised. Scientific disciplines are fields of research characterised by some form of consensus on theoretical assumptions, research questions, and relevant methodological tools. In addition, they are a community of scientists often organised in university departments that offer teaching in education programmes. These scientists publish in the same journals, visit the same conferences, and build on each other's research. Disciplines are important for science because they offer a community, a culture, and a way of doing things (Petts et al., 2008). In that sense, disciplines, quite literally, discipline scientists. They tell them what research problems to address, which theories and concepts to discuss, which methods to use, which styles of reasoning to apply to interpret and analyse results, and what to count as evidence for certain claims.

Currently, we distinguish a large number of different scientific disciplines, such as ecology, sociology, political science, soil science, physics, chemistry, psychology, and so on. However, within these main disciplines different and often partially overlapping sub-disciplines can be identified (clinical psychology, population ecology). These disciplines have a historical trajectory and themselves emerged as the result of collaboration between disciplines or the splitting of subject areas into two or more disciplines. For instance, ecology formed as a discipline in the 1960s, and split off from biology, and systems biology originated as a result of collaboration between specific sub-disciplines within biology, chemistry, and physics. There are many more examples. This means that disciplines, though very important in the daily work of scientists, are historically grown and culturally produced fluid entities with flexible boundaries that change over time (Lélé and Norgaard, 2005).

So, what is interdisciplinarity? In light of the above, we need to recognise that what we understand interdisciplinarity to be crucially depends on what we understand disciplines to be. This means that just like disciplines, interdisciplinarity itself is a rather unstable and contextual phenomenon that has many different meanings and interpretations. In this book, we understand interdisciplinarity to mean the collaboration between different scientific disciplines that are relevant for the problem at hand. In the case of environmental problems, we suggest that an interdisciplinary approach involves not just different natural science disciplines, but also social science disciplines.

The second part of the answer to the question is related to the 'inter' part of the term. While it is now clear that more than one discipline is involved in interdisciplinary work, there are different ideas about how knowledge from different disciplines can be combined. In what is often called 'multidisciplinarity' a certain topic or problem is addressed with multiple disciplines, but the insights generated by these disciplines are additive to each other and they are not integrated. The term 'interdisciplinarity' is used when knowledge from different disciplines is integrated. One other term that is increasingly used is transdisciplinarity. Transdisciplinarity involves collaboration beyond disciplines, and it also includes knowledge sources outside academia (of stakeholders, citizens, or professionals). We will discuss this in more detail in Chapter 8.

In practice, the boundaries between multi- and interdisciplinarity are often not clear because there are very different ideas about what 'integration' means and how it can be done. In this book, we will use the term 'knowledge integration' for those attempts to connect different forms of natural and social science knowledge under – and this is crucial – some kind of common framework or methodological approach. Before we offer some examples of these frameworks and approaches, the next section discusses the main barriers to knowledge integration.

7.3 Barriers to Knowledge Integration

While integrating knowledge is considered important, it is also often perceived as difficult. When considering our earlier discussion of disciplines, it is clear that scientists form communities with shared norms and values about what constitutes good science. It follows that crossing the boundaries between those communities, as is required for knowledge integration, can create problems: how is collaboration possible between scientists that hold different norms values for doing science? Lélé and Norgaard (2005) identify different barriers to knowledge integration that result from the differences between scientific communities or disciplines: values embedded in science, theories and explanatory models, epistemology, and institutional incentive structures. Below, we introduce those barriers in more detail.

As Chapter 2 also argues, all forms of scientific knowledge are influenced by choices about what to measure and how. These choices are not value neutral, since we tend to measure what we value – and we value what we measure. Thus, knowledge integration efforts may be hampered by 'mismatched taxonomies' (Lélé and Norgaard, 2005, 969) in which different scientists will put forward their preferred categories as candidates for knowledge production and integration. As a consequence, scientists in knowledge integration are confronted with the challenge to reflect on the values in science, something which natural scientists in particular are not used to doing, partly because they have been taught that (their) science is value neutral.

These values are related to theories and models of explanation. Particularly in complex problems, the relative superiority of one model or theory over another is difficult to establish, leading to problems in trying to decide on which to use in knowledge integration. Such decisions are also difficult because they are associated with fundamental, epistemological assumptions about what constitutes reliable knowledge. Differences between reductionist or systems approaches, between quantitative and qualitative methods, and between positivist and constructivist epistemologies will inevitably surface in interdisciplinary collaboration and are not easy to reconcile.

A different barrier refers to existing incentive structures in universities and other research organisations. Typically, scientific careers are based on disciplinary contribution and recognition, resulting in a general lack of attention for and appreciation of knowledge integration activities. Thus, while the problem at hand may require interdisciplinary collaboration, experts may not be inclined to prioritise this if it does not lead to papers that can be published in research journals that are held in high regard within the discipline.

7.4 Approaches to Knowledge Integration

Differences between disciplines and barriers to integration can be overcome – but this is, as we discuss in this section, not without costs – when a common framework for integration can be established. There are many different such frameworks. Some of these are quite close to the addition that takes place in multidisciplinary approaches, while others go far in integrating diverse forms of knowledge. In this section, we will discuss a number of approaches to knowledge integration. This is not intended as a full overview of methods and approaches, but serves to illustrate the variety of approaches.

7.4.1 Models and Scenarios

Among the first and most prominent examples of knowledge integration in the environmental domain is the study by the Club of Rome called 'Limits to growth: a global challenge' (Meadows et al., 1972). This study integrated knowledge from a wide variety of disciplines about e.g. population growth, industrialisation, pollution, agriculture, and resource use. The objective was to model relationships between economic and demographic trends and resource depletion, and to determine scenarios for the future. In particular, the business-as-usual scenario had a big impact on policy because it showed that in the near future, resources would become completely depleted. This study is an early example of an integrated assessment modelling study: a quantitative form of knowledge integration that uses standardised and common metrics as the framework for integration. The case about air pollution modelling (Case G: Integrated Assessment for Long-Range Transboundary Air Pollution) provides an example of an integrated assessment model in the environmental domain.

Integrated assessment modelling and scenario analysis are common tools to integrate knowledge. The models and scenarios used by the Intergovernmental Panel for Climate Change (IPCC) are a good example (Moss et al., 2010). The models are based on a number of theoretical assumptions about how the earth system and the economic system behave, which translate into complex algorithms that can be fed by data. They are then used to project future carbon emissions and their economic and environmental consequences relative to a number of scenarios. Each scenario is an alternative future situation, often called a storyline, which is then translated into assumptions about major environmental and socio-economic drivers of emissions.

A third example of model-based integration within the environmental domain is agent-based modelling. These models aim to predict the behaviour of actors or groups of actors using theoretical assumptions derived from, amongst others, rational actor theory, principal–agent theory, and the theory of planned behaviour. These can then be linked with other models to predict the effect of the predicted behaviour change on the environment (Bakker et al., 2015).

The framework for integration in modelling and scenario analysis is that they require that all data take the form of numbers (quantification), which, through calculation on the basis of theoretical assumptions, can ultimately be expressed in relevant units such as CO_2 and degrees Celsius that can be made equivalent to each other. For example, increased CO_2 emissions can be expressed in terms of euros, and rises in temperature can be expressed in terms of economic decline or human lives lost.

Scenario analysis and modelling approaches to knowledge integration can be combined with participatory approaches (Voinov and Bousquet, 2010; Kok et al., 2011). For example, a participatory scenario analysis may start by using narrative methods in which stakeholder groups are asked to develop and describe qualitative images of potential or desirable futures. Subsequently, these can be used as normative orientations or goals. When these goals are translated into quantitative expressions of a set of indicators that serve as a proxy for these goals, they can be linked to quantitative information or predictions about the trends in these indicators. The final step is to use this information to assess the likelihood of achieving these goals under different scenarios and to design measures to increase this likelihood. In participatory system dynamics modelling, participants can use models to compare different policy options and alternatives. Beall and Zeoli (2008) describe an example of such an approach in which participatory modelling was used to develop and evaluate land-use and management options for the conservation of a specific grouse species.

7.4.2 Multi-Criteria and Cost–Benefit Analysis

Decision makers often need integrated knowledge to support their decisions. They can use decision support tools that will help them to identify and balance the different expected consequences of certain plans or decisions. One of the common tools is cost–benefit analysis (CBA). CBA is a tool in policy planning that is used to balance the costs and benefits of specific projects and interventions and to come to an informed decision about whether or not to proceed with the plan (Hansjürgens, 2004). For example, they are recommended by the EU and widely used by EU member states in the implementation of EU environmental directives (Feuillette et al., 2015). For CBA to be possible, both the costs of a project and its benefits have to be expressed in a common, monetary unit; in other words, the data have to be fully commensurate. For the application of CBA to environmental issues, this is not a simple matter since environmental costs and benefits also need to be included. How to translate expected changes in nature and environment into monetary units? There are many different economic valuation methods that can be used (Hanley and Spash, 1993). However, they all have limitations, and there is considerable discussion about what items in nature and environment should be valued, and whether and how monetary value should be assigned to these items (Wegner and Pascual, 2011; Bartkowski et al., 2015).

A second common decision support tool is multi-criteria analysis (MCA). MCA differs from CBA in two important ways. First, it allows for the inclusion of multiple criteria, while CBA only focuses on monetary costs and benefits. Second, although MCA does require commensuration of the criteria since all values must be expressed quantitatively, it does not assign absolute values in a standardised unit like CBA does. Instead, MCA enables the ranking of alternative options, expressed as unitless numbers, against a set of weighed criteria (Gamper and Turcanu, 2007). The following simple example explains the basic process: imagine a policy maker who wants to reduce traffic congestion in her village. She can consider two options: 1) broaden the existing road through the centre of the village; or 2) build a ring road around the village. The first step in doing the MCA is to identify the criteria that will be used to assess the options. In this example these can be public health, costs, and nature conservation. The second step is to identify indicators for each criterion. For public health, for example, you can use air quality, and for nature conservation you can use habitat loss. Next, you will need knowledge about the effects of the two options. For example, you can calculate the costs of the two options, the expected reduction in air pollution in the village centre, and the loss of habitat in the forest surrounding the village where the ring road is planned. In this simple example, the MCA may produce the following scores for the different criteria. The option to broaden the village road scores low on public health and high on costs and nature conservation, while the ring road option scores high on public health but low on costs and nature conservation. A very important step in an MCA is deciding the relative importance of each criterion. Depending on the weight of the criteria, the options can be ranked. If all criteria are ranked equally, the option to broaden the village road is preferred, but if public health is given more weight, the ring road may become the preferred option.

Most MCAs are much more complicated than this example. They will use more than two options, include more than three criteria with multiple indicators for each criterion, use complicated calculations and models to assess the impacts of the options, and allow for a wider range of scores than simply high or low. Most applications of MCA are, to a varying extent, participatory, with participants identifying the criteria and indicators and deciding on how to weigh the different criteria (see Sheppard and Meitner, 2005, for an example in forest management). Thus, MCA is a way to balance and compare things that are not directly commensurable. By using numbers to facilitate this balancing and comparing, the problem of the incommensurablility of values and criteria is partly overcome, but also hidden from view.

7.4.3 Conceptual Frameworks

Another way of linking different forms of quantitative and qualitative informa-
tion is through a so-called conceptual framework. A conceptual framework offers a succinct overview, using pictures or words, of the key social and ecological variables that are to be included, as well as the processes through which these variables are linked and how they are to be understood. As such, a conceptual framework communicates how different aspects of the issue at stake in the assessment are related, and, therefore, how different forms of knowledge about these aspects can be linked. They offer a common structure and a

common terminology that allows scientists from different disciplines to collaborate and to understand each other's contribution.

Within the environmental domain, the concept of ecosystem services is
increasingly used for these purposes. By making explicit the links between
the environment, the goods and services humans derive from the environment,
and the contribution of those goods and services to human well-being, the
concept offers an integrated framework that couples social and ecological
systems. This is one reason why the concept was central to the conceptual
framework used in the Millennium Ecosystem Assessment (Case F:
Knowledge Integration in the Millennium Ecosystem Assessment).

Apart from this function of offering a representation of social–ecological
systems, the concept of ecosystem services is also considered attractive
because it emphasises the value of nature and biodiversity for people and for
the economy. Yet, this strong economic and utilitarian focus has also resulted in
considerable criticism – specifically, the way in which it frames the problem of
conservation as an economic problem to be solved by means of market
mechanisms, thereby ignoring the intrinsic value of nature and the different
ways in which humans relate to and value nature that go beyond economic
transactions has been the focus of considerable critical discussion (Muradian et
al., 2013; Turnhout et al., 2013).

Figure 7.1 presents the conceptual framework used by the Intergovernmental
Platform for Biodiversity and Ecosystem Services (IPBES) as their model for
integration. In this conceptual framework, the concept of ecosystem services
takes centre stage. However, the idea of the authors is that it works like a
Rosetta Stone (Díaz et al., 2015) which allows different actors, within and
outside science, who have very different ways of understanding nature and
environment, to engage in meaningful conversation and dialogue, and which
facilitates mutual translation of knowledge between those who employ the
vocabulary of 'mother nature' and 'nature's gifts' and those who use terms
such as 'biodiversity' and 'ecosystem services' (Borie and Hulme, 2015).

Forms of knowledge integration that are guided by a conceptual framework are less formal than integration in models or CBA and do not require
the full commensuration of knowledge. It is possible that different quantitative and qualitative methods and tools will be used to assess the different
elements in the conceptual framework. The same applies to the connections between the elements: these can be assessed by means of modelling
and scenario analysis, but also by means of plausible stories and other
qualitative approaches. The advantage of this form of knowledge integration
is that it builds on the strengths of different disciplines and does not, for
example, impose quantification or single metrics on social sciences. This

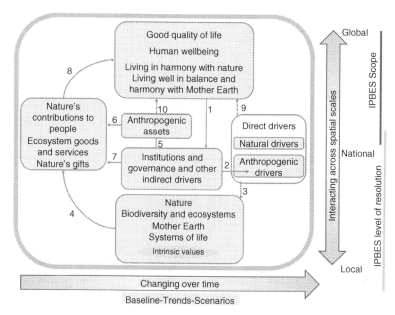

Figure 7.1 The conceptual framework of the Intergovernmental Platform for Biodiversity and Ecosystem Services (adapted from Díaz et al., 2015).

immediately also signals what can be seen as a weakness of this approach. Formal integration remains limited as the different forms of knowledge that get integrated are not comparable and may be based on different understandings of the elements of the conceptual framework.

The examples discussed in this section illustrate the different possibilities in methods and approaches for knowledge integration. Specifically, we have seen important differences when it comes to how and to what extent they integrate. For some methods, integration involves full commensuration – the expression of all knowledge in a single standardised quantitative measure – while some are based on quantification but do not require a single measure, and others offer a common language or conceptual framework to guide integration.

7.5 The Politics of Knowledge Integration

What we have seen in the previous section is that different approaches differ with respect to the type and extent of integration. In quantitative approaches, different forms of knowledge are all expressed as numbers and are made commensurable to enable integration in a literal or formal sense. In more loose or qualitative approaches, integration is perhaps closer to what we could call aggregation or addition. However, as is also clearly explained in the case about the Millennium Ecosystem Assessment (Case F: Knowledge Integration in the Millennium Ecosystem Assessment), there are often good reasons for not quantifying different forms of knowledge and for not striving for full commensuration. As we explained earlier, different scientific disciplines have different ways of working, and the social sciences and the humanities in particular have long and rich traditions in qualitative approaches. To translate this qualitative knowledge into numbers cannot be done without sacrificing some of the meaning and quality of that work (Breslow, 2015). This is also true for lay expertise (see Chapter 8).

This is why the idea of knowledge integration is problematic. Quantification and commensuration may allow for a higher degree of integration, but the knowledge that is subjected to quantification and commensuration changes fundamentally. Rather than bringing in the wealth of diverse knowledge, this diversity is reduced to the lowest common denominator, removing the bits that do not match instead of seeing them as an opportunity to learn. There is also an irony to consider here: if such a high degree of integration requires such intensive processes of translation of knowledge from different disciplines, to what extent can the product then still be seen as interdisciplinary? Rather, you could say, this is a new knowledge production process that is itself structured by its own 'monodisciplinary' rules and methods. Paradoxically, then, this form of knowledge integration ends up reducing the disciplinary diversity that is considered necessary for dealing with complex problems and that drove the impetus for knowledge integration in the first place.

Looser, or more qualitative forms of knowledge integration, such as those that use a conceptual framework, may not conform to some of the ambitions of knowledge integration, particularly the creation of a single comprehensive assessment that has the authority to compel action. Yet, they are potentially more successful in connecting science, policy, and society because they are able to facilitate dialogue or what Jasanoff (1998) has called 'reasoning together'. To be sure, more closed and quantitative approaches to knowledge integration can also play a role in facilitating dialogue, particularly when they are combined with participation (Peterson et al., 2003; Breslow, 2015). This is also the main lesson from the case on air pollution modelling (Case G: Integrated Assessment for Long-Range Transboundary Air Pollution): the modelling was seen as useful because it facilitated learning and dialogue and, arguably, this was more important in compelling action in the implementation of the agreement than the scientific quality of the modelling results. However, there is an important difference in the type of dialogue that can take place. Since commensuration erases and reduces differences between forms of knowledge, it in effect also depoliticises issues by suggesting that all knowledge can be included, that common ground can be found, and that fully integrated knowledge can inform rational decisions that will benefit everybody. Conceptual frameworks that go less far in the commensuration of knowledge can facilitate a more political form of dialogue that allows for differences to be recognised and addressed explicitly.

The point of the chapter has not necessarily been to choose between different approaches, but to highlight advantages and disadvantages, to emphasise the importance of knowledge integration processes as opportunities for learning and interaction, and to reflect on the political implications of different knowledge integration approaches. We suggest that knowledge integration efforts should take care to safeguard the diversity and plurality of knowledge systems, recognise the value of having a wide variety of mono-, inter-, and multidisciplinary approaches, and carefully consider the risks and opportunities of different approaches to knowledge integration.

References

Bakker, M. M., Alam, S. J., Van Dijk, J., and Rounsevell, M. D. A. (2015). Land-use Change Arising from Rural Land Exchange: An Agent-Based Simulation Model. *Landscape Ecology*, 30(2), 273–286.

Bammer, G. (2005). Integration and Implementation Sciences: Building a New Specialization. *Ecology and Society*, 10(2), art. 6.

Bartkowski, B., Lienhoop, N., and Hansjürgens, B. (2015). Capturing the Complexity of Biodiversity: A Critical Review of Economic Valuation Studies of Biological Diversity. *Ecological Economics*, 113, 1–14.

Beall, A., and Zeoli, L. (2008). Participatory Modelling of Endangered Wildlife Systems: Simulating the Sage-Grouse and Land Use in Central Washington. *Ecological Economics*, 68, 24–33.

Borie, M., and Hulme, M. (2015). Framing Global Biodiversity: IPBES between Mother Earth and Ecosystem Services. *Environmental Science & Policy*, 54, 487–496.

Breslow, S. J., (2015). Accounting for Neoliberalism: 'Social Drivers' in Environmental Management. *Marine Policy*, 61, 420–429.

Díaz, S., Demissew, S., Joly, C., Lonsdale, W., and Larigauderie, A. (2015). A Rosetta Stone for Nature's Benefits to People. *PLOS Biology*, 13(1), p.e1002040

Feuillette, S., Levrel, H., Boeuf, B., et al. (2015). The Use of Cost–Benefit Analysis in Environmental Policies: Some Issues Raised by the Water Framework Directive Implementation in France. *Environmental Science & Policy*, 57, 79–85.

Gamper, C., and Turcanu, C. (2007). On the Governmental Use of Multi-criteria Analysis. *Ecological Economics*, 62(2), 298–307.

Hanley, N., and Spash, C. (1993). *Cost–Benefit Analysis and the Environment*. Aldershot: Edward Elgar.

Hansjürgens, B. (2004). Economic Valuation through Cost-Benefit Analysis? Possibilities and Limitations. *Toxicology*, 205(3), 241–252.

Jasanoff, S. (1998). Harmonization: The Politics of Reasoning Together. In R. Bal and W. Halfman, eds., *The Politics of Chemical Risk* (pp. 173–194). Dordrecht: Kluwer Academic Publishers.

Kok, K., Van Vliet, M., Bärlund, I., Dubel, A., and Sendzimir, J. (2011). Combining Participative Backcasting and Exploratory Scenario Development: Experiences from the SCENES Project. *Technological Forecasting and Social Change*, 78(5), 835–851.

Lélé, S., and Norgaard, R. (2005). Practicing Interdisciplinarity. *BioScience*, 55(11), 967–975.

Meadows, D. H., Meadows, D. L., Randers, J., and Behrens, W. W. (1972). *The Limits to Growth*. New York: Universe Books.

Miller, C. A. (2007). Democratization, International Knowledge Institutions, and Global Governance. *Governance: An International Journal of Policy, Administration, and Institutions*, 20, 325–357.

Mitchell, R. B., Clark, W. C., Cash, D. W., and Dickson, N. M., eds. (2006). *Global Environmental Assessment: Information and Influence*. Cambridge: The MIT Press.

Moss, R., Edmonds, J., Hibbard, K., et al. (2010). The Next Generation of Scenarios for Climate Change Research and Assessment. *Nature*, 463(7282), 747–756.

Muradian, R., Arsel, M., Pellegrini, L., et al. (2013). Payments for Ecosystem Services and the Fatal Attraction of Win-Win Solutions. *Conservation Letters*, 6, 274–279.

Petts, J., Owens, S., and Bulkeley, H. (2008). Crossing Boundaries: Interdisciplinarity in the Context of Urban Environments. *Geoforum*, 39(2), 593–601.

Peterson, G., Cumming, G., and Carpenter, S. (2003). Scenario Planning: A Tool for Conservation in an Uncertain World. *Conservation Biology*, 17(2), 358–366.

Sheppard, S. R. J., and Meitner, M. (2005). Using Multi-criteria Analysis and Visualisation for Sustainable Forest Management Planning with Stakeholder Groups. *Forest Ecology and Management*, 207, 171–187.

Turnhout, E., Waterton, C., Neves, K., and Buizer, M. (2013). Rethinking Biodiversity: From Goods and Services to 'Living With'. *Conservation Letters*, 6, 154–161.

Voinov, A., and Bousquet, F. (2010). Modelling with Stakeholders. *Environmental Modelling & Software*, 2010, 25, 1268–1281.

Wegner, G., and Pascual, U. (2011). Cost–Benefit Analysis in the Context of Ecosystem Services for Human Well-being: A Multidisciplinary Critique. *Global Environmental Change*, 21(2), 492–504.

Case F Knowledge Integration in the Millennium Ecosystem Assessment

CLARK MILLER

The Millennium Ecosystem Assessment is an example of an integrated assessment study. The study is well known because of its use of the now common but also contested concept of Ecosystem Services. The case illustrates the advantages and challenges of integrating different forms of knowledge, including natural sciences, social sciences, and indigenous and local knowledge using a conceptual framework (Section 7.4) and it also illustrates the barriers to and political implications of knowledge integration (Sections 7.3 and 7.5).

Introduction

The Millennium Ecosystem Assessment, or Millennium Assessment (MA), offers a unique case study of the problem of knowledge integration. Officially launched in 2001 and published in 2005, the MA sought to create a scientific foundation for global biodiversity policy. Institutionally, the MA built on and expanded the model of the Intergovernmental Panel on Climate Change (IPCC), which many ecologists viewed at the time as a highly successful effort to apply knowledge to international governance. From the beginning, however, the MA faced a different set of knowledge integration challenges than the IPCC. The MA responses to these challenges illuminate some of the key epistemic choices that face institutions confronting global problems in the twenty-first century, and so are worth examining in detail.

The core of the MA, like the IPCC, was organised around a small number of international working groups tasked with assessing the state of scientific knowledge about biological diversity and ecosystems. The science of biodiversity and ecosystems differs in key ways, however, from the science of climate change. Since the 1980s, climate science has revolved around a handful of computational models of the Earth's climate system – models that simulate global-scale biogeophysical processes and coordinate climate science activity on both a conceptual and an organisational basis (see, e.g., Shackley and Wynne, 1995; Edwards, 2010). While a few rough parallels exist in the science of biological diversity, such as efforts to measure the global number and distribution of species (e.g. Myers, 1979; Olson and Dinerstein, 1998; Doubilet et al., 1999; Mittermeier et al., 1999) or to model global ecological systems (Yue et al., 2011), these projects do not play the same orchestrating role in the field of biodiversity as their counterparts do in climate science. Rather, the vast majority of theoretical, empirical, and applied work in ecology and biodiversity conservation has occurred on very local scales. In fact, the primary ecological research tool arguably remains the field plot, while the focus of ecological studies for most scientists remains on the dynamics of localised facets of the world's natural environment: organisms, communities, ecosystems, niches, and landscapes (Levin et al., 2009).

At the same time, the governance of biodiversity also differs markedly from that of climate change, starting with problem framing. In the case of climate change, the Earth's climate system is viewed as global (Miller, 2004), as are the markets for fossil fuels and the companies that extract, process, and sell them (Perkins, 2004). The major drivers of change are likewise located in the heart of the global economy, in the world's largest (and richest) countries, tied centrally to the organisation of global capital (Huber, 2013; Mitchell, 2013; Klein, 2015). By contrast, nations generally view biodiversity in terms of biological or natural resources – whether in terms of landscapes, e.g. forests, fisheries, agriculture, or in terms of genetics (on the latter, see, e.g., Correa, 1995; Diaz, 2005) – to be under their sovereign control (Schrijver, 2008), with many of the world's important regions of biological diversity located in developing countries (Mittermeier et al., 1999).

From the beginning, therefore, knowledge integration in the MA was organised differently from the IPCC. Three aspects of this difference are particularly significant. First, whereas participants in the IPCC took for granted its conceptual foundations as rooted in scientific models of the Earth's climate system (Hulme, 2010), the organisers of the MA conducted an explicit exercise to draft a common conceptual framework as the first step of the assessment (Alcamo et al., 2003) prior to forming its scientific working groups. In building this framework, they not only sought to articulate an explicit narrative within which to frame an assessment of global biodiversity, they also sought to do so on an interdisciplinary basis. Second, the MA's scientific working groups included much more extensive and effective participation from scientists from developing countries (Reid et al., 2006) than the IPCC had accomplished (Kandlikar and Sagar, 1999; Ho-Lem et al., 2011). Finally, while the IPCC considered itself as solely a global assessment, the organisers of the MA approached their work as a multi-scale assessment, carrying out a top-down, global assessment while also catalysing the creation of a highly diverse network of bottom-up, sub-global assessments (Miller and Erickson, 2006). What follows is organised around these three aspects of MA knowledge integration.

Knowledge Integration and the MA Conceptual Framework

In the late 1990s, the World Resources Institute, a prominent environmental non-governmental organisation (NGO) with a deep interest in biodiversity conservation, began to explore with several international organisations the formation of a global scientific assessment of biological diversity. From the outset, these discussions identified effective communication as a key challenge for biodiversity science. While global climate negotiations had achieved what seemed at the time to be a landmark agreement in 1997 in the Kyoto Protocol, parallel negotiations under the Convention on Biological Diversity seemed to have stalled in the face of several profound geopolitical conflicts. Motivated by what they viewed as a profound global crisis of biodiversity destruction, the World Resources Institute interpreted the lack of movement towards a new global biodiversity agreement as a clarion call for renewed efforts to advance the voice of science in ways that would persuasively portray the significance of biodiversity to world leaders and thus motivate rapid policy action (Takacs, 1996; Rosen, 2000).

When the World Resources Institute and the United Nations launched the MA in 2001, organisers focused immediately on the development of a compelling conceptual framework and narrative structure for the assessment. Frustrated by their lack of policy impact, ecologists had worked throughout the 1990s on several efforts to create analytic methods that could frame ecological degradation in more human terms. These efforts came together in the mid-1990s around two key concepts: ecosystem services and socio-ecological systems. In both cases, ecologists worked with social scientists to build integrated conceptual models. The idea of ecosystem services sought to integrate ecological and economic analyses to reframe ecology as a problem of natural resources on which human societies depended for critical services (Daily, 1997; Robert et al., 1998). The idea of socio-ecological systems, in turn, sought to recognise that social and ecological systems were interdependent on one another in a variety of complex ways that could only be fully understood and analysed in an integrated fashion (Berkes et al., 1998).

Building on both of these ideas, the MA conceptual framework was developed in a series of meetings in 2002. The framework identified ecosystem services as the core conceptual foundation for the assessment while also recognising that these services influence, and are influenced by, the dynamics of broader social, economic, political, and technological processes and systems at multiple scales. This conceptual effort, captured in Figure F1, was seen as so important by both the organisers and the biodiversity conservation community

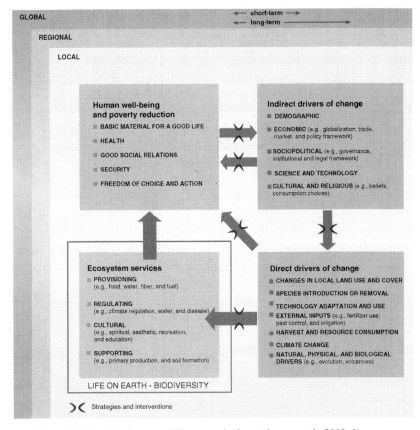

Figure F1 The MA Conceptual Framework (from Alcamo et al., 2003, 9).

that its results were not only circulated to all participants in the assessment as the basis for their work, but also quickly published as the MA's first report *Ecosystems and Human Well Being* (Alcamo et al., 2003), and became one of the MA's most visible, iconic, and highly cited products.

The MA conceptual framework is remarkable for the diversity of knowledge traditions represented even in the highly simplified diagram in Figure F1. Represented are numerous major policy domains and their specialised knowledges (poverty, health, security, food, water), as well as key social sciences (economics, political science, anthropology, sociology, and science and technology studies), numerous facets of interdisciplinary environmental and agroeconomic studies (land use, invasive species, climate change), and underlying ecological processes. At the same time, the framework is especially significant for its explicit recognition of the importance of narrative in knowledge integration. Knowledge integration is often viewed as a technical problem of aggregating and piecing together diverse scientific facts, data, and theories. In this view of knowledge integration, the central challenge is the fragmentation of science. In terms of work practice, individual scientists and laboratories study only small parts of complex natural processes and systems, and thus generate partial and fragmented knowledge. Complicating the picture further, scientists tend to work in disciplines whose assumptions and models structure inquiry in limiting ways. The task of assessments is to piece the resulting fragments together to reassemble the whole picture.

What MA leaders recognised – as scholars had also begun to argue (see e.g. Cronon, 1992) – is that knowledge integration also has a larger narrative dimension. In that sense, the MA operated on the working assumption that there is no privileged, universal position from which to assemble the facts of nature. Rather, just as the analysis of data in science is theory-laden, so, too, the integration of knowledge in assessments takes shape within larger storylines that structure both the ways facts are assembled and synthesised and the social meanings they express. The MA leadership thus viewed the conceptual framework as more than a mere tool for organising the assessment. They saw it as an instrument for motivating, standardising, and ultimately rationalising world-wide policy efforts to conserve biodiversity.

It is also worth noting, however, that the MA conceptual framework was only partially successful, even on its own terms. Participants and others have criticised the framework for limiting the assessment's scope in ways that prevented it from grappling with the full complexity of the social and ecological processes at play in biodiversity conservation, arguing that assessments of this kind can never fully pull together fragmentary disciplinary knowledge (e.g. Norgaard, 2008, 2010). Others have lamented that the MA framework was too broad, failing to tame the inevitable localisation of biodiversity conservation and, thus, to create the kind of clear, global scientific foundation for policy action provided by the IPCC (Loreau et al., 2006). Finally, as we will recount in this case, the MA framework was at best marginally successful in providing a common foundation for the work of the MA sub-global assessments, which adopted a remarkable diversity of epistemological foundations for understanding the relationships between humans and ecosystems around the globe.

Integrating North–South

A second major element of knowledge integration in the MA involved efforts to pull together perspectives from scholars in both the global North and South.

The preceding major scientific assessment of biodiversity, *The Global Biodiversity Assessment* (Watson, 1996), had seen its influence sunk, single-handedly, by the government of Malaysia. Upon its first presentation to the Conference of Parties of the Convention on Biological Diversity, the Global Biodiversity Assessment had been met with a sceptical Malaysian response that governments had not asked for such an assessment, at which point considera-tion of its findings was permanently tabled. This history convinced MA organ-isers that the world's developing countries must play a major role in both authorising and carrying out the assessment. Ultimately, the main offices for the MA were located in the World Fish Center in Penang, and major efforts were made to ensure extensive participation of scientists from developing countries at all levels of the organisation, from the governing board and scientific leadership team to the scientific working groups and chapter author teams. Many developing countries also organised sub-global assessments.

The success of knowledge integration involving North–South participants varied considerably across different facets of the assessment. Two of the MA scientific working groups, *Current State and Trends* and *Policy Responses*, ultimately proved relatively straightforward in terms of integration, although for different reasons. The *Current State and Trends* working group was tasked with examining the existing status of the world's ecosystems and the direction of trends in that status. This formulation offered a congenial framing for ecologists from all parts of the world and working in all different kinds of ecosystems to document their knowledge as part of an aggregation of factual knowledge about global ecology. Differences in the scales, epistemologies, geographies, and contexts adopted by individual researchers seemed not to particularly matter as they worked together to tally up what was known about the state of the world's ecological systems. The *Policy Responses* working group struggled from the outset to bridge the differences between key intellec-tual perspectives, such as development economics and ecological economics. In the end, however, the disciplinary strength and conventions of economics within the team facilitated the creation of a broadly shared framework. Co-leadership of the working group by a very strong team in India also ensured that the framework did not slight perspectives from the world's poorer regions.

By contrast, knowledge integration was less successful in the *Scenarios* working group. Much as their counterparts in the IPCC have done, scientists working within the MA approached scenario development largely as a project for computational modelling of global systems. Since little such modelling is done in developing countries, this significantly limited the potential for core scientific participation in the group by scientists from poorer regions of the world. Not surprisingly, therefore, the scenarios developed largely reflected established themes within the North about environment, development, and biodiversity conservation – in stark contrast to some of the outputs from the bottom-up-organised sub-global assessments. The *Scenarios* working group did put significant effort into connecting these scenarios to dynamics in developing countries, interviewing leaders from many parts of the globe as part of their initial background research, incorporating perspectives from India and South Africa in the depiction of each scenario, and working with the MA leadership, building a major outreach and engagement strategy involving local interpretation of the scenarios in diverse parts of the world.

Bridging Scales and Epistemologies in the MA Sub-Global Assessments

The MA's third and most ambitious effort at knowledge integration came via its commitment not only to build a global assessment of biodiversity knowledge, but also to foster the development of a large number of sub-global assessments. This initiative ultimately resulted in the creation of more than 30 local, regional, and sub-global assessments across the globe, all of which adapted the MA conceptual framework to their own specific ecological and policy contexts. The ultimate goal of this initiative was to build a second, parallel assessment alongside the global MA but from the bottom up, drawing out of these myriad assessments a more comprehensive, multi-scale picture of the world's ecological systems rooted in on-the-ground knowledge, integrated into regional and ultimately global syntheses.

In the end, the sub-global initiative was a mixed success. MA leaders recognised three key factors from the outset that positioned the MA differently from the IPCC: (1) that most ecologists and other researchers working on biodiversity conservation produce knowledge primarily at local or regional scales; (2) that ecology, society, politics, and markets operate in an interlinked fashion across scales; and (3) that conducting a global assessment alone would both do gross injustice to the epistemic foundations of biodiversity conservation and fail to engage a key set of local and regional decision makers who would continue to have enormous influence over the future of the planet's biological diversity. Yet, the MA leadership also felt strongly that the absence of a global perspective and analysis was harming biodiversity conservation.

To resolve this dilemma, the MA fashioned a multi-scale assessment framework that could integrate local, regional, and global knowledge and perspectives (Capistrano, 2005). Their aim was to ensure that local and regional knowledge informed the global assessment, and vice versa, improving decision making at all scales. In tackling this effort, the MA also acknowledged that a central problem for multi-scale assessment would be not simply be to integrate knowledge across scales, but also to bridge distinct epistemologies. As already observed, this was already a challenge in the interdisciplinary context of the global assessment, which sought to integrate knowledge from ecology with economics. Yet, the multi-scale assessments went further, attempting to bridge not only multiple disciplines, but also diverse national approaches to risk and environmental assessment, as well as to incorporate indigenous and scientific knowledge (Reid et al., 2006; Miller and Erickson, 2006).

To carry out its work, the MA multi-scale assessment commissioned more than 30 'sub-global' assessments, a phrase chosen to reflect the remarkable diversity of its efforts, which ranged from a single city park to Sub-Saharan Africa. Each sub-global assessment was asked to ground its activities in the MA conceptual framework, but unlike other global assessments that have sought to build in multi-scale insights, the MA sub-global assessments were allowed to develop their own metrics and methodologies – and this, too, emerged as a major element of diversity. To illustrate this diversity, in the largest sub-global assessment a group of scientists from South Africa coordinated a continent-wide ecological analysis of Sub-Saharan Africa. Working with collaborators in numerous countries, this assessment followed the full MA model, documenting conditions and trends, examining future scenarios, and detailing policy response options. At the other end of the scale, in Peru, a small team of ecologists partnered with indigenous and local communities to conduct an indigenous assessment of local ecological trends and conditions in the Vilcanota region, producing a fascinating set of resources built on indigenous cultural and spiritual insights.

Integrating these assessments into the MA proved difficult. For one, the management of dozens of diverse assessments by local groups proved challenging within the framework of the global MA initiative. As the MA's three global working groups were launching their efforts in 2002, applications for sub-global assessments were just beginning to be solicited and evaluated, and local teams were just being pulled together. MA funding provided only partial support for most of the MA sub-global assessments, which meant local organisers took more time to secure funding from local sources. Thus, the initial idea that the sub-global assessments would provide knowledge inputs into the global assessment largely ran afoul of timing problems. Many were not even complete at the time the global MA was finalising its report for publication in 2005. At the same time, the heterogeneity of the sub-global assessments also prompted challenges for mundane tasks such as data integration and standardisation that would be required to fully incorporate sub-global insights into global knowledge-making.

The epistemic diversity of the MA sub-global assessments also made it difficult to pull them together as a group to say something meaningful, in global terms, about local and regional ecosystems. As already noted, while some of the assessments were relatively straightforward ecological assessments, others drew heavily on indigenous knowledge traditions. Even among the ecological assessments, problem framings varied enormously, as did the underlying theoretical foundations of the problem. While the MA conceptual framework imposed a certainly loose similarity of language across all of the assessments, it couldn't paper over the epistemological differences between assessments conducted, e.g. of a city park, based on an intellectual foundation rich in socio-ecological systems theory; an assessment of slash-and-burn agriculture across dozens of locations scattered around the equator; and a sub-continent scale effort to conduct a scaled-down version of the global MA (including ecological status and trends, scenarios, and policy responses assessments) across scales, from local to regional, for Sub-Saharan Africa.

What emerged from the sub-global assessment working group and its report was thus not a full synthesis of the knowledge from the diverse sub-global assessments, in the sense that no serious attempt was made to try to pull the diverse assessments together into a global picture of ecosystems. Rather, what emerged was a modest form of what Jasanoff (1998) terms 'reasoning together': a strategy for recognising that how people in different parts of the world reason about the problems they face is culturally situated and grounded in different epistemologies. Even when the problems confronting forms of governance and its reliance on science appear similar, diverse epistemologies of regulatory science can evolve with very different styles of reasoning, norms of evidence, and definitions of expertise (Jasanoff, 2005). Bridging these differences, Jasanoff (1998) argues, requires finding approaches that allow people to learn to reason together: to confront one another's forms of rationality, learn why they differ, acknowledge their strengths and limitations, and develop new, shared languages for making sense of the world.

What the MA sub-global assessment thus hinted at is that this form of knowledge integration is the most difficult and time consuming, but also the most important in confronting global challenges. Even though it failed to achieve anything like a full knowledge synthesis, the MA sub-global assessment initiative should be seen as a good attempt at knowledge integration. The challenge facing the world is not merely that most of the world's scientific knowledge is organised by research projects that are local, regional, or national in scope and scale (and thus need to be synthesised to tell us a global story).

Rather, the challenge is also that people in different parts of the world reason differently – grounded, for example, in different civic epistemologies (Miller, 2008) – about how to understand and act in response to changes in the global environment. To successfully bridge these differences, knowledge integration needs to be pursued differently than mere aggregation and synthesis of data. As suggested by the MA, it will be a challenge of working out of collectively shared ways of seeing, imagining, and reasoning about global problems.

References

Alcamo, J., Hassan, R., Scholes, et al. (2003). *Ecosystems and Human Well-being.* Washington, DC: Island Press.

Berkes, F., Folke, C., and Colding, J. (1998). *Linking Social and Ecological Systems: Management Practices and Social Mechanisms for Building Resilience.* Cambridge: Cambridge University Press.

Capistrano, D. (2005). *Ecosystems and Human Well-being: Findings of the Sub-global Assessments Working Group.* Washington, DC: Island Press.

Correa, C. (1995). Sovereign and Property Rights over Plant Genetic Resources. *Agriculture and Human Values,* 12(4), 58–79.

Cronon, W. (1992). A Place for Stories: Nature, History, and Narrative. *The Journal of American History,* 78(4), 1347–1376.

Daily, G. (1997). *Nature's Services.* Washington, DC: Island Press.

Díaz, C. (2005). *Intellectual Property Rights and Biological Resources: An Overview of Key Issues and Current Debates.* Wuppertal: Wuppertal Institute for Climate, Environment and Energy.

Doubilet, D., Grove, N., Lanting, F., and Rowell, G. (1999). *World Wildlife Fund Presents: Living Planet.* New York: Crown Publications.

Edwards, P. (2010). *A Vast Machine: Computer Models, Climate Data, and the Politics of Global Warming.* Cambridge: MIT Press.

Ho-Lem, C., Zerriffi, H., and Kandlikar, M. (2011). Who Participates in the Intergovernmental Panel on Climate Change and Why: A Quantitative Assessment of the National Representation of Authors in the Intergovernmental Panel on Climate Change. *Global Environmental Change,* 21(4), 1308–1317.

Huber, M. (2013). *Lifeblood: Oil, Freedom, and the Forces of Capital.* Minneapolis: University of Minnesota Press.

Hulme, M. (2010). Problems with Making and Governing Global Kinds of Knowledge. *Global Environmental Change,* 20(4), 558–564.

Jasanoff, S. (2005). *Designs on Nature, Science and Democracy in Europe and the United States.* Princeton: Princeton University Press.

Jasanoff, S. (1998) Harmonization: The Politics of Reasoning Together. In R. Bal and W. Halffman, eds., *The Politics of Chemical Risk* (pp. 173–194). Dordrecht: Kluwer Academic Publishers.

Kandlikar, M., and Sagar, A. (1999). Climate Change Research and Analysis in India: An Integrated Assessment of a South–North Divide. *Global Environmental Change*, 9(2), 119–138.

Klein, N. (2015). *This Changes Everything*. London: Penguin Books.

Levin, S., Carpenter, S., Godfray, H., et al. (2009). *The Princeton Guide to Ecology*. Princeton: Princeton University Press.

Loreau, M., Oteng-Yeboah, A., Arroyo, M., et al. (2006). Diversity Without Representation. *Nature*, 442(7100), 245–246.

Miller, C. (2004). Climate Science and the Making of a Global Political Order. In S. Jasanoff, ed., *States of Knowledge: The Co-Production of Science and the Social Order* (pp. 46–66). London: Routledge.

Miller, C. (2008). Civic Epistemologies: Constituting Knowledge and Order in Political Communities. *Sociology Compass*, 2(6), 1896–1919.

Miller, C., and Erickson, P. (2006). The Politics of Bridging Scales and Epistemologies: Science and Democracy in Global Environmental Governance. In W. Reid, F. Berkes, and T. Wilbanks, eds., *Bridging Scales and Knowledge Systems* (pp. 297, 314). Washington: Island Press.

Mitchell, T. (2013). *Political Power in the Age of Oil*. London: Verso Books.

Mittermeier, R., Goettsch Mittermeier, C., Myers, N., and Robles Gil, P. (1999). *Biodiversidad Amenazada*. Mexico: CEMEX.

Myers, N. (1979). *The Sinking Ark*. Oxford: Pergamon Press.

Norgaard, R. (2008). Finding Hope in the Millennium Ecosystem Assessment. *Conservation Biology*, 22(4), 862–869.

Norgaard, R. (2010). Ecosystem Services: From Eye-Opening Metaphor to Complexity Blinder. *Ecological Economics*, 69(6), 1219–1227.

Olson, D., and Dinerstein, E. (1998). The Global 200: A Representation Approach to Conserving the Earth's Most Biologically Valuable Ecoregions. *Conservation Biology*, 12(3), 502–515.

Perkins, J. (2004). *Confessions of an Economic Hit Man*. San Francisco: Berrett-Koehler Publishers.

Reid, W., Capistrano, D., Berkes, F., and Wilbanks, T. (2006). *Bridging Scales and Knowledge Systems*. Washington: Island Press.

Robert, C., d'Arge, R., de Groot, R., et al. (1998). The Value of the World's Ecosystem Services and Natural Capital. *Ecological Economics*, 25, 3–15.

Rosen, C. (2000). *World Resources 2000–2001*. Amsterdam: Elsevier.

Schrijver, N. (2008). *Sovereignty over Natural Resources*. Cambridge: Cambridge University Press.

Shackley, S., and Wynne, B. (1995). Integrating Knowledges for Climate Change: Pyramids, Nets and Uncertainties. *Global Environmental Change*, 5(2), 113–126.

Takacs, D. (1996). *The Idea of Biodiversity*. Baltimore: Johns Hopkins University Press.

Watson, R. (1996). *Global Biodiversity Assessment*. New York: Cambridge University Press.

Yue, T., Jorgensen, S., and Larocque, G. (2011). Progress in Global Ecological Modelling. *Ecological Modelling*, 222(14), 2172–2177.

Case G Integrated Assessment for Long-Range Transboundary Air Pollution

WILLEMIJN TUINSTRA

This case on activities under the Convention On Long-Range Transboundary Air Pollution (CLRTAP) is an illustration of a simultaneous process of integrating knowledge and of integrating science and policy. It shows how during the process of data collection and negotiation, scientists and policy makers together frame a problem, determine what aspects are important, and define relevant and workable indicators. The use of the RAINS model offers an example of a model used as a common framework for integration, as presented in the main text of Chapter 7. The case shows that the logic of integration that was used led to an expanding network of people, issues, and tools. At the same time it shows that there is a limit to comprehensiveness.

The Convention on Long-Range Transboundary Air Pollution

The 1970s and 1980s witnessed significant developments in international cooperation to deal with environmental and health issues. In 1972, the Stockholm Declaration of the United Nations Conference on the Human Environment stated that nations have 'the responsibility to ensure that activities within their jurisdiction or control do not cause damage to the environment of other States or of areas beyond the limits of national jurisdiction'. In the same year, the United Nations Environment Programme (UNEP) was created. A few years later, the 1975 Helsinki Conference on Security and Cooperation in Europe called for cooperation to control air pollution and its effects, including long-range transport of air pollutants. In 1979, 30 countries adopted the Convention on Long-range Trans-boundary Air Pollution (CLRTAP). This convention established a general forum for international negotiations on emission reductions of air pollutants in which not only western and eastern European countries participated, including the Soviet Union, but also the Unites States and Canada (these were still Cold War times). In 1983, the parties signed a protocol to finance a Cooperative Program for Monitoring and Evaluation of the Long-Range Transmission of Air Pollutants in Europe (EMEP). Later, a Working Group on Effects and a Task Force on Integrated Assessment Modelling were formed to enable the development of effect-oriented and cost-effective policy strategies. Thus, a framework for monitoring and scientific assessment of air pollution was established in terms that were decidedly transnational. In addition, the ambitions were not just to assess pollution effects across national borders, but also across different types of pollution, providing an excellent example of knowledge integration and its problems.

In this case study, we will highlight the activities that were initiated in CLRTAP to integrate and create knowledge to deal with the problems associated with long-range transboundary air pollution. These included acidification, eutrophication, and ground-level ozone, but also the health effects of particulate matter and the interlinkages between air pollution and climate change. Negotiations on various emission reduction agreements (protocols) coincided with a huge international effort to collect, validate, and streamline scientific information and integrate data on every aspect connected to air pollution, ranging from data on soil conditions, weather patterns, and deposition and concentrations of pollutants, to estimates of emissions, control costs of those emissions, and the identification and reduction potential of available technologies. There was a strong drive to integrate all these different aspects, motivated by exactly the rationales mentioned in Chapter 7: 1) the urge to create a comprehensive picture that covers 'all relevant aspects' of the problem; 2) the need for an integrated assessment of the potential effectiveness of various policy options to solve the problems; and 3) the need to include various sources of knowledge in order to enhance the legitimacy of solutions and the effectiveness of their implementation. The 'various sources of knowledge' were understood to involve the countries and the data they collected and provided (as opposed to one single team of experts deriving the data for the countries). A related, but slightly different aspect of integration in this case was the strong cooperation between scientists and policy makers. During the process of data-collection and negotiation, scientists and policy makers together framed the problem, determined what aspects were important, defined relevant and workable indicators, defined an understanding of when the problem could be considered solved, and defined the relevant actors and sources for information. As we will see, this was a dynamic process in which the outcomes of these negotiations (e.g. relevant indicators or ideas about relevant actors and knowledge) could change the framing of the integration project.

Harmonising Approaches

Typically, the first protocol under the Convention was not a protocol about the reduction of a specific pollutant, but about ensuring finances for information collection: the EMEP protocol mentioned above. The EMEP had three main components: collection of emission data for sulphur dioxide (SO_2), nitrogen oxides (NO_x), volatile organic compounds (VOCs), and other air pollutants; measurement of air and precipitation quality; and modelling of atmospheric dispersion. At present, more than 200 monitoring stations in 40 UNECE countries continue to participate in the programme. From the beginning, CLRTAP established a dual framework of scientific assessment and political interaction, which gradually changed, but still functions. A number of working groups and task forces regularly reviewed the scientific information in the fields of environmental impacts of air pollution, atmospheric dispersion, emission control technologies, economic evaluation methods, and integration methodologies. Also, databases with quality-controlled information about the country-specific situations were prepared. The outcome of this scientific assessment process was submitted to the Working Group on Strategies and Review, whose members are civil servants representing governments of parties to the Convention, which used this information to assist its negotiations on further emission control agreements. Hence, finances and essential protocols in this cooperation had a strong focus on gathering and integrating knowledge.

In the course of ca. 30 years between 1985 and 2015, integration increasingly took shape. The Second Sulphur protocol (the 'Oslo Protocol'), signed in 1994, was the first protocol that established a quantitative relation between the agreed emission reduction targets and environmental impacts. This reflected a wider trend in environmental policy towards effect-oriented policy: increasingly, the focus was no longer on pollution or emissions as such, but on the environmental effects of these emissions (Hajer, 1995; Turnhout et al., 2007). In this new form of policy, new types of knowledge and expertise were required. Earlier protocols under the Convention (for reducing emissions of SO_2 (1985), NO_x (1988), and VOC (1991)) had used simple 'flat rate' concepts for determining the international distribution of reduction obligations, i.e. all countries agreed to reduce their emissions by the same percentage, or stabilise emissions in the case of NO_x, relative to a base year. In contrast, the reduction obligations of the Oslo Protocol were based on the results of modelled linkages between the SO_2 emissions of each country and the environmental effects;

the exposure of different ecosystems to deposition, taking into account the sensitivity of such ecosystems to acidification, as well as atmospheric transport; and the regional differences in costs of emission reduction. As a consequence, the most cost-effective distribution of efforts resulted in differences in reduction obligations among countries. The 'Gothenburg-protocol' (1999) was a next step in integration in the sense of creating a comprehensive picture of the problem. This multi-pollutant/multi-effect protocol covered SO_2, NO_x, NH_3, and non-methane volatile organic compounds, and aimed at the simultaneous reduction of acidification, eutrophication, and ground-level ozone. Like in the Oslo Protocol, reduction obligations differentiated among the countries. Such a differentiation only had been possible through 1) the availability of the comprehensive data sets collected in the international networks, 2) the creation of an over-arching integrative concept for the scientific assessment, the so-called 'critical loads', and 3) the development and use of computer modelling tools (Integrated Assessment Models) in order to combine and analyse all information in a consistent and efficient way. Knowledge integration provided a platform for negotiation and bargaining over a wide range of issues and policy fields, thus leading to policy integration. Particularly important had been the political acceptance to use this simplified way to quantify ecosystem risks in negotiations.

With regard to *data-collection* (e.g. for the assessment of atmospheric dispersion), the EMEP, with its meteorological synthesising centres and chemical coordinating centres in Oslo and Moscow, used (and uses) official emission inventories provided by national governments and evaluated by the Task Force on Emission Inventories (Lövblad et al., 2004). In this way, the EMEP was able to relate deposition to emissions in each European country. The collection and mapping of data on critical loads was arranged by the Coordinating Centre for Effects, which was established at the National Institute for Public Health and Environment (RIVM) in the Netherlands. In all countries party to the Convention, national focal centres were established to provide this data. A unified methodology was developed in order to ensure international harmonisation. Various initiatives were taken to further harmonisation, including the development of a mapping manual, workshops with specialists from all over Europe, and expert training sessions. The mapping work was an iterative process and the maps were regularly updated (Hettelingh et al., 2015). Hence, not only concepts and environmental processes were made comparable, but data was also made commensurate to make integration possible.

Integrated Assessment Modelling and the Concept
of Critical Loads

Critical loads have been defined as 'the maximum exposure to one or several pollutants, at which according to current knowledge no harmful effects occur to sensitive ecosystems in the long run' (Nilsson and Grennfelt, 1988). In the absence of fully quantified ecological damage functions, critical loads were simplifications representing a steady-state 'no-damage' threshold. With the use of critical loads, the information collected by the Working Group on Effects was made usable for the development of policy options and measures. This was further facilitated by the acceptance of the 'cost-effectiveness' principle that was used to identify options that could lead to an efficient reduction in critical load exceedance. The combined approach of the principle of critical loads as well as of cost-effectiveness was instrumental in negotiating the environmental ambition level and the resulting emission reduction requirements for the Oslo Protocol of 1994. Several authors have pointed to the importance of the critical loads approach as a common basis for air pollution control within both the policy and the science communities (e.g. Hajer, 1995; Bäckstrand, 2017; Sundqvist et al., 2002). Its success is attributed to the capacity 'to create a meaningful framework and to invoke common sense notions of adequate approaches to environmental policy-making' (Bäckstrand, 2001, 136).

All collected information was integrated in the *Integrated Assessment Models* (IAMs) that were used by the Task Force on Integrated Assessment Modelling (TFIAM) of CLRTAP. This task force was established in 1986, before the actual negotiations on the Oslo Protocol began (Tuinstra et al., 2006). Its task was 'to explore the possibilities to develop an analytical framework for a regional cost-benefit and cost-effectiveness analysis of concerted policies to control air pollution' (UNECE, 1986). At that moment, several models were under development at various research institutes in Europe. TFIAM discussed and compared various model approaches and made suggestions for further development. The final calculations that served as a starting point for the negotiations of the Oslo Protocol, as well as for the Gothenburg protocol, were done by RAINS, the model of the International Institute for Applied Systems Analysis (IIASA) (Tuinstra et al., 2006). RAINS is a modular model that connects economic activities with emissions (and their control policies and costs), the dispersion of pollutants through the atmosphere, and the environmental effects of these pollutants (Alcamo et al., 1990; Amann et al., 1996). This last effect-module uses the concept of critical loads to assess the effects of projected depositions of acidifying and eutrophying pollutants.

Currently, the IIASA at Laxenburg, Austria, functions as the Conventions' Centre for Integrated Assessment Modelling (CIAM) and centrally uses the GAINS model, an extended version of the RAINS model (Amann et al., 2011). CIAM organises so-called bilateral consultation meetings with all parties to agree upon national input data and to gain credibility from the model output. In recent years, the modelling work has been extended to include, amongst others, the health risks of air pollution on the local level (particulate matter), and to incorporate the relation between air pollution and climate change (Reis et al., 2012). Other changes have also been made, including the release of an online version of GAINS. These extensions and changes, just like the early expansions with the economic module and the possibility to compare scenarios in the 1980s, have all been done in anticipation of, or in response to, new needs of negotiators and other policy makers.

Conclusion

The use of the RAINS model offers an example of a model used as a common framework for integration, as presented in Chapter 7. It is a quantitative approach to integration, which requires that all data take the form of numbers, which, through calculation on the basis of theoretical assumptions, can ultimately be expressed in relevant units that can be related to each other (control costs, critical loads). Or, put differently, the concept of critical loads was able to function as a boundary object (Star and Griesemer, 1989) that was flexible enough to connect actors with different perspectives and stable enough to provide sufficient common ground for cooperation (also see Chapter 6).

This case shows that the logic of integration that was used led to an expanding network of people, issues, and tools. Not only RAINS/GAINS as a model and the work of TFIAM, but also LRTAP as a process extended over the course of the years: more pollutants were incorporated, more effects were taken into account, and more environmental issues were integrated. At the same time, more countries were involved, more experts contributed, and also cooperation with other organisations and negotiating arenas were established. For example, LRTAPs Scientific Assessment report of 2016 advised on a shift towards more attention for black carbon. LRTAP cooperated on this with the Arctic Monitoring and Assessment Programme (AMAP) of the Arctic Council. Furthermore, in 2016 an UNECE/WHO/UNEP interagency cooperation on transboundary air pollution was set up. The idea was to join efforts to complement and strengthen each other's work and to be able to have a bigger impact by providing joint messages to governments.

At the same time it is important to note that there is a limit to comprehensiveness, and, as we have shown with framing and climate models (see Chapter 2 and Case A: Framing Climate Change), 'the' full picture does not exist. Scientists involved in the data-collection and modelling processes are aware that 'more complete' models do not automatically lead to better decisions. Yet the commitment to intensified integration continues in the belief that in any case integration of nitrogen policy, climate policy, and air policy is necessary in order to streamline measures and decisions about them. In the years to come, TFIAM will work on further integration of scales (local, regional, and global air pollution) (Maas and Grennfelt, 2016). Awareness of the limits of harmonisation as well as targeted efforts to remain transparent will remain crucial for the effectiveness for this further integration. Transparency is all the more important when more people and organisations are involved. Clearly, LRTAP is open to new frames of problems and to new groups of actors who might have different relevant questions. Also, the tools used are flexible. What has made them effective in the past is the insight of the people involved that not only the tools themselves but also the processes that were put in place to foster data collection, international cooperation, and networking were as important. Therefore, probably their most important contribution was that they fostered learning and dialogue. The critical loads concept and the integrated assessment modelling tools served to bring data and networks of people together.

References

Alcamo J., Shaw R., and Hordijk L., eds. (1990). *The RAINS Model of Acidification. Science and Strategies in Europe*. Dordrecht: Kluwer Academic Publishers.

Amann M., Bertok, I., Cofala, J., et al. (1996). *Cost-effective Control of Acidification and Ground-Level Ozone*. Laxenburg: International Institute for Applied Systems Analysis.

Amann, M., Bertok, I., Borken-Kleefeld, J., et al. (2011). Cost-effective Control of Air Quality and Greenhouse Gases in Europe: Modelling and Policy Applications. *Environmental Modelling and Software*, 26, 1489–1501. doi:10.1016/j. envsoft.2011.07.012

Bäckstrand, K. (2001). *What Can Nature Withstand? Science, Politics and Discourses in Transboundary Air Pollution Diplomacy*. Lund Political Studies 116. Department of Political Science. Lund University.

Bäckstrand, K. (2017). Critical Loads: Negotiating What Nature Can Withstand. In J. Meadowcroft and D. J. Fiorino, eds., *Conceptual Innovation in Environmental Policy* (pp. 129–154). Cambridge, MA: MIT Press.

Hajer, M. A (1995). *The Politics of Environmental Discourse: Ecological Modernization and the Policy Process*. Oxford: Oxford University Press.

Hettelingh, J. P., Posch, M., Slootweg, J., et al. (2015). Effects-based Integrated Assessment Modelling for the Support of European Air Pollution Abatement Policies. In W. de Vries, J. P. Hettelingh, and M. Posch, eds., *Critical Loads and Dynamic Risk Assessments of Nitrogen, Acidity and Metals for Terrestrial and Aquatic Ecosystems* (pp. 613–635). Dordrecht: Springer. DOI:10.1007/978–94–017–9508–1_25

Lövblad, G., Tarrason, L., Tørseth, K., and Dutchak, S. (2004). *EMEP Assessment, Part I, European Perspective.* Oslo: Norwegian Meteorological Institute.

Maas, R., and Grennfelt, P., eds. (2016). *Towards Cleaner Air. Scientific Assessment Report 2016.* EMEP Steering Body and Working Group on Effects of the Convention on Long-Range Transboundary Air Pollution. Oslo.

Nilsson, J., and Grennfelt, P., eds. (1988). *Critical Loads for Sulphur and Nitrogen.* The Nordic Council of Ministers 15. Copenhagen.

Reis, S., Grennfelt, P., Klimont, Z., et al. (2012). From Acid Rain to Climate Change. *Science,* 338, 1153–1154.

Star, S. L., and Griesemer, J. (1989). Institutional Ecology, 'Translations', and Boundary Objects: Amateurs and Professionals in Berkeley's Museum of Vertebrate Zoology, 1907–1939. *Social Studies of Science,* 19, 387–420.

Sundqvist, G., Letell, M., and Lidskog, R. (2002). Science and Policy in Air Pollution Abatement Strategies. *Environmental Science & Policy,* 5(2), 147–156.

Tuinstra, W., Hordijk, L., and Kroeze, C. (2006). Moving Boundaries in Transboundary Air Pollution Co-production of Science and Policy under the Convention on Long Range Transboundary Air Pollution. *Global Environmental Change,* 16(4), 349–363.

Turnhout, E., Hisschemöller, M., and Eijsackers, H. (2007). Ecological Indicators: Between the Two Fires of Science and Policy. *Ecological Indicators,* 7(2), 215–228.

UNECE (1986). Report 4th EB-meeting 1986: ECE/EB.AIR/10. Geneva: United Nations.

8

Lay Expertise

ESTHER TURNHOUT AND KATJA NEVES

8.1 Introduction

Lay expertise is a term that can be used to typify non-scientific forms of
knowledge. Alternative terms one may find in the literature are lay knowledge,
local knowledge, folk knowledge, traditional ecological knowledge, and indigenous knowledge. Lay expertise is recognised as important for conservation
and for sustainable resource management in both Western and non-Western
contexts. Yet it is also often considered inferior to science, and this hampers its
inclusion in conservation and management practice. In this chapter, we will
discuss the contribution of lay expertise, how it can be harnessed effectively,
and how participatory approaches can be used in knowledge production.
We will do this by advancing a symmetrical perspective that focuses on
commonalities and complementarities between different scientific and non-
scientific forms of knowledge while at the same recognising and respecting
difference and diversity.

Lay expertise is often defined as the opposite of scientific knowledge because
it is considered to be: contextual and localised rather than universal; culturally
embedded rather than objective; tacit and informal rather than explicit and
formalised; practice and experience based rather than based upon methodological principles; methodology based; intuitive rather than cognitive; and holistic
rather than reductionist (Gadgil et al., 1993; Berkes, 1999). However, as we have
seen in Chapter 2, the question of what science is and how it can be distinguished
or demarcated from other forms of knowledge is far from self-evident. And this
also holds for the demarcation between lay expertise and science.

Nevertheless, this binary system to classify and distinguish lay expertise
from scientific expertise is prevalent in discussion about lay expertise. Its main
effect is that it leads to hierarchical comparison that can either render invisible
the crucial insight that lay expertise often offers or privilege it as a more truthful

form of knowledge. As a consequence, some consider lay expertise to be inferior to science because of its defining characteristics, while others deem these same characteristics to be more valuable, thus constituting a superior form of knowledge when compared to science.

Increasingly, scholars are pointing to the complementarity of scientific and lay expertise, the productive synergies they may embody (Neves-Graca, 2004; Collins and Evans, 2008; Ponting and O'Brien, 2014). Also, it is becoming increasingly common to see examples of interaction and collaboration between different kinds of actors bearing different kinds of knowledges (see, for example, Case F: Knowledge Integration in the Millennium Ecosystem Assessment), including those that carry the labels of 'science' and 'lay expertise' (Lane et al., 2011; Whatmore and Landström, 2011; Tsouvalis and Waterton, 2012). Before we turn to these opportunities for collaboration, we first discuss the value of lay expertise.

8.2 The Value of Lay Expertise

Studies exploring the value of lay expertise have recently become common in domains regarding land use, including issues of cultivation, gaming and hunting, recreation, nature management, and nature restoration. One important objective of these studies, which often used labels such as 'traditional ecological knowledge' or 'local ecological knowledge', was to demonstrate the value of non-scientific ways of knowing. Specifically, they argue that lay expertise can promote social goods such as livelihoods, environmental protection, and biodiversity conservation (Gadgil et al., 1993; Berkes et al., 2000; Herrera et al., 2014; Downie, 2015). One theory argues that societies that hold knowledge of an ecosystem's diversity and relationships are not only likely to value these 'resources', but also often provide important insights into preservation practices and their rationale (Milton, 2002; Wall, 2014; Pahl-Wostl, 2015). Another theory postulates that communities that manage their resources collectively and that are informed by lay expertise are able to avoid the destructive forms of economic extraction that stem from individualistic and competitive economic behaviour (Neves-Graca, 2004; Yang, 2015). However, some have warned against these conceptions of lay knowledge and community-based management, which are often founded on romanticised notions of local peoples as 'noble savages' whose environmental impact is always benign and/or of little consequence (Redford, 1991; Edgerton, 1992; Alvard, 1993; Ellingson, 2001). Indeed, numerous examples of instances where lay expertise failed to prevent environmental disaster exist (Acheson, 2006). Moreover, it has been shown that an ethic of conservation or sustainability does not always emerge from common property ownership (Ruttan, 1998; Neves, 2010).

Studies advocating the importance of lay expertise also often see science and modernisation as a threat. The centrality of science and technical expertise privileged systems of rationality and utilitarianism, associated with the development of modernity, colonialism, and capitalism (Latour, 1993; Nandy, 1988), to the detriment of a heterogeneity of socio-ecologically and religiously embedded knowledge systems (Gaukroger, 2006). According to Shiva (1993, 9), an advocate for lay expertise and a critic of science, modern scientific knowledge systems not only emerge 'from a dominating and colonising culture', they 'are themselves colonizing'. Thus, with the adoption of modern lifestyles, local cultures and their knowledges are increasingly threatened and lay expertise about nature and environment may become 'extinct' (Turnbull, 2009). An example of this is offered by Van der Ploeg, who shows how the introduction of modern agricultural techniques in the Andes resulted not just in disappointing harvests, but also in the deskilling of local farmers and the erasure of their expertise (Van der Ploeg, 2002).

In response, a number of initiatives have attempted to collect, archive, and, in that way, conserve such knowledge, such as the People's Biodiversity Register Program (Agrawal, 1995; Gadgil et al., 2000). Equally, there are initiatives to include lay expertise in global biodiversity assessments, as discussed in the case of the Millennium Assessment (Case F: Knowledge Integration in the Millennium Ecosystem Assessment). However, as we saw in that case, there are several challenges associated with these initiatives. One risk is that lay expertise is only evaluated from a scientific perspective, and that it can only be used when it is translated into forms that are compatible with science. Such a scientisation of lay expertise reflects the uneven power relations that exist between science and lay expertise and fails to recognise the multiple cultural and contextual dimensions of lay expertise (Bryan, 2011).

While it is generally recognised that interactions between different forms of knowledge may produce positive outcomes for knowledge, for conservation and for other actors in society, it is a question of who benefits from lay expertise and how these benefits are distributed. Several examples of what is often called biopiracy show that dominant actors have been able to appropriate lay expertise and valuable natural resources without proper acknowledgement of intellectual property rights or compensation (Shiva, 1997; Kloppenburg, 2010; see Case H: Lay Expertise and Botanical Science, for an example).

The Convention on Biological Diversity (CBD) has incorporated measures to ensure equal access and benefit sharing, and its rules prescribe the inclusion and the protection of lay expertise. These measures are put in place to prevent the appropriation of lay expertise and of the benefits and profits that may be derived from it. This formal recognition of lay expertise makes clear that lay expertise is potentially a politically charged resource that can be used to advocate the needs of its holders. One example is indigenous co-management of natural resources. Co-management arrangements, for example in the Arctic North, are required to respect and incorporate lay expertise (the co-management literature mostly uses the term 'traditional ecological knowledge') in the negotiation of management practices, including the rights of indigenous people to access and harvest resources by indigenous people (Nadasdy, 1999, 2003).

As discussed earlier in this volume, it has been important for science to claim its authority by showing how it is different from, and better than, other forms of knowledge. This chapter shows that advocates of lay expertise have used a very similar strategy. To argue for the value of lay expertise, they had to show that it is a special form of knowledge and that it is different from other forms of knowledge, including science. However, as the remainder of this chapter will argue, strict boundaries between forms of knowledge neglect the multiple interactions between them and the considerable variations within the categories of scientific and lay expertise (also see Agrawal, 1995; Turnbull, 1997; Goldman, 2007).

8.3 What Happens When Different Forms of Knowledge Meet?

In practice, interactions between science and lay expertise are dynamic and complex, including not just exclusion and polarisation, but also complementarities. We will give four examples.

Wynne (1996) describes an example of sheep farmers in the English Lake District in the aftermath of the 1986 accident at the nuclear power plant in Chernobyl, Ukraine. In the days after the accident, a north-western wind brought clouds carrying radioactive caesium over Europe and the UK. Two main areas in the uplands of the UK – the Lake District in Cumbria and the mountains of North Wales – were directly affected by the accident as they were subject to torrential rain, which deposited the radionuclides over the high fells. Following this, the main question at stake for British politicians was whether it was safe for the sheep to graze on the fells or whether the levels of radioactive contamination were too high. For farmers, this question was of crucial importance: if the contamination was too high, the sheep could no longer be marketed for human consumption and would need to be slaughtered. Based on model studies of standardised rates of decay of radionuclides such as radio caesium, scientists projected that the levels would drop rapidly and therefore a temporary, three-week ban on fell grazing would be appropriate. However, these projections included considerable uncertainties, which the scientists failed to mention. For example, they did not take into account the chemical composition of Cumbrian peaty acid soils, which differed greatly from the alkaline soils on which the model projections were based. The projections were shown to be inaccurate when field studies failed to report decreasing contamination levels. As the scientists had not adequately communicated the uncertainties involved in the projections and did not admit to their mistakes, farmers had every reason to distrust these scientific experts. Patronising experts seemed to fit a pattern, as locals remembered how experts had also played down the risks of another nuclear accident: the fire in 1957 that destroyed the nuclear power reactor at Windscale, on the West Cumbrian coast. This sense of growing distrust increased still further when experts gave unrealistic advice to the farmers: experts told the farmers to keep their sheep on the lower parts of the fells and in the valleys where contamination was lowest and to supplement their diet with straw. This advice completely disregarded farming practices and sheep behaviour (sheep cannot be contained locally while grazing, and they do not eat straw). The point of this example is that a lack of trust in scientific knowledge and expertise was not based on a lack of understanding of science. On the contrary, the farmers possessed considerable knowledge: about the past behaviour of scientific experts not communicating uncertainties and not admitting to mistakes; about the incorrect projections and advice they produced; and about the local situations and farming practices that the scientific experts had failed to take into account. Thus, they had highly rational reasons to distrust science, and they had relevant expertise to offer themselves.

The Chipko movement, which started in early 1970s in India, offers a second example of interactions between lay and scientific expertise (Rangan, 1996). The movement started in the form of protests by local people from the eastern districts of Garhwal Himalayas who tried to stop the felling of trees by hugging them. The protests were based on lay expertise: the local people of Chipko knew that their marginalisation and poverty were related to deforestation and forest degradation, which caused problems of soil degradation, erosion, and floods. Through these protests, they tried to stop commercial logging and they criticised the local government for failing to take effective conservation action. While the movement started as dispersed local protests, educated elites – including scientists, activists, or combinations of the two – increased the legitimacy of the protests by promoting them as a form of grassroots environmentalism and by further substantiating their claims. In that way, the movement became a hybrid mix of scientific and lay expertise and gained attention from national and international audiences. The demands made by the Chipko movement were recognised by the Indian government, who issued a 15-year logging ban (Shiva and Bandyopadhyay, 1986). However, there is some controversy about the success of the movement as some authors argue that the protests resulted in increased government control and restricted local access to the forest (Brown, 2014).

A third example of interactions between local and scientific forms of knowledge is natural history and biodiversity recording or monitoring. A significant portion of the world's biodiversity data is collected by local people and volunteers (Ellis and Waterton, 2004; Lawrence, 2010; Toogood, 2013; Bruijnikx, 2015). Formal institutions for biodiversity science and conservation, including natural history museums, rely to a great extent on the data delivered by volunteers (Star and Griesemer, 1989; Shirk et al., 2012). Through their participation in biodiversity monitoring, lay experts not only contribute to science, but also enact their personal relationship with their environment (Lawrence and Turnhout, 2010). Despite continued questioning of the reliability of volunteer data (Gura, 2013; Turnhout et al., 2016), amateur naturalists are often accepted as authoritative biodiversity experts. As such, natural history is a domain of inquiry with blurry boundaries between scientific and lay expertise, and with multiple collaborations and interactions between scientists and volunteers.

The transition that took place in the archipelago of the Azores (Portugal) from whale hunting to whale watching offers the fourth example of the complementarity that can exist between lay expertise and scientific knowledge (Neves-Graca, 2004, 2006). This became evident in 1999 when efforts were made to regulate whale watching in order to avoid the likely negative impact of heavy boat loads near whales and dolphins, and the associated disturbances that boat engines cause to these mammals. While in general everyone agreed that creating this legislation should be a priority, disagreement quickly ensued between a group of former whale hunters who had become whale-watching entrepreneurs and a group of whale-watching business owners allied with marine biologists. The disagreement between the two groups stemmed from different understandings of the kinds of whale-watching practices that are detrimental to whales and dolphins and the kinds that are not. Whale-watchers working with marine scientists relied on generalised knowledge of whale responses to whale watching, often obtained from analysing data obtained from afar or pertaining to a wide variety of whale and dolphin species in many areas of the planet. They were convinced that whale watching was generally safe for whales, insofar as proper distances and careful approaches are respected. The former whale hunters, in turn, relied on intimate knowledge of sperm whales and their behaviour within the specificities of the marine ecosystems of the Azores. They argued that extra measures to protect whales from whale-watching harassment ought to be implemented given that many whales nurse their offspring in those waters and that the high-pitched noise produced by off-board engines can frighten them. The former whalers were well versed in the biology and science of whales: in addition to the knowledge they had accumulated for more than a century, they had also learnt a great deal from some of the best marine scientists in the world, who had spent time in the Azores and to take advantage of whale-hunting activities in order to study their biology. Not surprisingly, the boundaries between the whale hunter's lay expertise and the scientific knowledge of the marine scientists were rather fluid. While the former whale hunters faced major struggles in getting their voices heard, both groups came to realise that their knowledges of whales and the potential effects of whale watching were in fact complimentary. The legislative package that now regulates whale watching in the Azores was the outcome of this complementarity.

The four examples offered here show that there are no universally valid boundaries between science and lay expertise, and that the stereotypes discussed in the first section of this chapter cannot be upheld. First, the examples show that science is not always superior to lay expertise, and that lay expertise can be equally valid, or true. Second, the examples show that scientific and lay forms of expertise are not necessarily incompatible; they can be used together to make sense of specific environmental phenomena. Even more so, the mutual influences of science and lay expertise can have beneficial outcomes as the cases about botanical science (Case H: Lay Expertise and Botanical Science) and Loweswater (Case I: The Loweswater Care Project) show. Thus, the conventional categories of scientific knowledge and lay expertise are too static, they ignore similarities and historical interdependencies, and they exaggerate differences. To better account for these overlaps and complementarities, we propose an alternative symmetrical conception of knowledge as situated practice that applies to scientific and lay forms of expertise.

8.4 Knowledge as Situated Practice

The discussion of lay expertise so far has made clear that knowledge production processes are not restricted to science but take place in society as well, in a large variety of settings, including scientific expert fora, local communities, and civil and professional organisations. While we do not assume that these knowledges are 'the same', we do believe that they cannot be defined using simple binaries and categories such as 'science' or 'lay expertise'. Thus, what is required is a conceptualisation of knowledge and expertise that is able to appreciate the provisional, social, and non-essential character of knowledge systems; their differences and their similarities. The concept of 'situatedness' is crucial to this conceptualisation. The idea of situated knowledge states that all knowledge, be it scientific or lay, is partial, generated in specific contexts in social processes, and its validity is assessed in equally contextual and social processes (Haraway, 1988). From that perspective, science emerges not as the exclusive way to produce truthful knowledge, but as a site in which a special kind of localised and contextualised knowledge is produced while making claims to universal validity and truthfulness. This context-specificity of science is clearly demonstrated by the different traditions and styles in the production and use of expertise, and the different ideas about what constitutes relevant and valid knowledge in different countries, political systems, and regulatory fields (Halffman, 2005; Jasanoff, 2005).

Raffles (2002) offers a useful account of how scientific and lay expertise come about and intersect. He relates a series of research experiments that were carried out in the Amazon to investigate the length of the roots of Amazonian trees and grasses. Dominant ecological theory posited that Amazonian trees had shallow roots and that nutrient cycling was a closed system. The research assistant involved in setting up the experiments admits: 'I knew some of these trees [had deep roots]. I've always known . . . that there were big trees and some vines with a long . . . root that brought up water from way below the surface' (Raffles, 2002, 326). Thus, when the experiments revealed the presence of deep roots, this aligned well with the lay expertise of the assistant, but not with dominant scientific theories. To be scientifically credible, this knowledge had to become codified not just by means of experiments, but also in the subsequent writing up. As Raffles (2002, 327) explains:

> Though the particular . . . local knowledge may hold the secret to scientific progress (it may guide us to those deep roots for example, in the same scientific terms it nevertheless lack universalism. To become meaningful on a planetary scale, Moacyr's [the name of the research assistant] knowledge of the root system must be translated into a language of expertise, incorporated into and subsumed by the mobile narratives of natural science.

In other words, as this example makes clear, while all knowledge is situated, some knowledges become scientised through various steps of translation, while other knowledges do not. The situatedness of knowledge, and the conceptual-isation of the adjectives 'scientific', 'lay', or 'local' as the outcome of processes of translation, interpretation, and articulation, suggest that differences are created rather than essentially given, and that they are not set in stone. Specifically, the move to recognise science as situated and as produced in local practice serves to create a more equal space for collaboration between forms of knowledge in which diversity and differences are respected but not seen as irreconcilable.

8.5 Participatory Knowledge Production

In the twentieth century, several developments resulted in the emergence of new ideas about how to organise the production of knowledge and the relation between science and society. Several people, including scientists (Moore, 2009), started to question the priorities of research, the strong connections between science, industry, and particular forms of modern or industrial agriculture, and the negative social effects of science. In conjunction with broad-based calls for the democratisation of a wide variety of government institutions and new demands for participation in decision-making processes, notions of democratising science and knowledge production have been brought to the fore (Chilvers and Kearnes, 2015).

As part of this democratisation agenda, it was considered important to bridge the perceived gap between science and citizens by using participatory approaches in the production of knowledge. These arguments align well with the perspective on lay expertise outlined in this chapter. If, on the level of philosophy or epistemology, there is no way to a priori demarcate between science and lay expertise, and if the relevance of lay expertise is increasingly considered important, the logical step is to include lay expertise in processes of knowledge production. Particularly in complex environmental problems, which involve high scientific uncertainties, it has been argued that decisions must be based on more than just scientific knowledge and expertise (Funtowicz and Ravetz, 1993).

Several academics have started to develop theories, concepts, and methodologies about participatory forms of knowledge production. One influential contribution has been the notion of Post-normal Science, which was developed in the 1990s by Funtowicz and Ravetz (1993). They argue that normal science is not able to adequately address current social and environmental problems and the high uncertainties and risks they involve. For these problems, science cannot offer definitive answers, and, instead, what is required is a post-normal approach that involves non-scientists in the production and evaluation of knowledge (Funtowicz and Ravetz, 1993).

A second influential concept is transdisciplinarity. Transdisciplinary approaches go beyond conventional ideas of multi- or interdisciplinary collaboration (see Chapter 7). Rather, they *trans*cend (hence the term *trans*-disciplinarity) the idea of discipline and consider knowledge production as a participatory and deliberative process wherein all different forms of knowledge are included on equal terms. A key aspect of transdisciplinarity is that it is problem oriented, starting from real-world problems and focusing on the collaborative development of robust solutions (Gibbons et al., 1994; Klein et al., 2001).

The literature commonly distinguishes between three different reasons for participatory approaches in knowledge production: 1) normative reasons related to democratic rights and principles; 2) substantive reasons related to the expected higher quality of knowledge; and 3) instrumental reasons that highlight the increased uptake of knowledge and the effectiveness of resulting policy or management interventions (Stirling, 2006). These different reasons often co-exist, but there are also risks that too much emphasis on certain reasons may crowd out other reasons. If the main focus is on creating better knowledge, this may exclude knowledge that is not considered useful; or, if the reasons are primarily instrumental, the focus will be on including only those forms of knowledge that are easy to incorporate.

These challenges are also visible in an area of participatory knowledge production known as citizen science. Citizen science can be defined as the participation of citizens in science; yet, the employment of the concept reveals important differences. One interpretation of citizen science foregrounds instrumental reasons and focuses primarily on the capacity of citizens to collect data and contribute to science (Bonney et al., 2009). Natural history and biodiversity recording (see Section 8.3) are examples of this interpretation of citizen science. A second interpretation of citizen science is much more normative and focuses on how citizens can use science to further their own objectives and pursue environmental justice (Irwin, 1995; Haklay, 2013). One example of the latter are the so-called bucket brigades, in which citizens use cheap and accessible technologies – so-called buckets – to measure air pollution in order to advocate for clean-air policies (Ottinger, 2010). Citizen science is a burgeoning field, with a high number of initiatives and projects currently being undertaken, a specialised scientific journal, *Citizen Science: Theory and Practice*, and a number of national and regional associations across the globe. These associations have demonstrated awareness of the importance of citizen science to strike an appropriate balance between each of the three reasons mentioned earlier. For example, the European Citizen Science Association has adopted ten principles that emphasise the importance of equal participation, of meeting citizens' needs, and of reciprocal relations between scientists and citizens, that all citizen science projects should respect (https://ecsa.citizen-science.net).

As the discussion of citizen science has also demonstrated, the implementation of participatory knowledge production is complicated. This is partly because it requires different types of skills and expertise, including social skills related to involving stakeholders and organising and facilitating participatory processes (Turnhout et al., 2013; also see Chapter 9 of this volume for more on the different roles of science in society). A different – and more important – reason is that there is a significant power dimension to these processes that needs to be carefully attended to (Stirling, 2006, 2008). Science can be a tremendous source of power, and the participation of non-scientists is often constrained by a perceived lack of skills, education, and eloquence, or by the limited availability of time and resources (Pellizzoni, 2001). Finally, the scope for participation is often limited by preconceived ideas – or frames – regarding what the issue at stake is about, which kinds of knowledge are relevant, who should be included, and how the participants are expected to behave (as lay persons in need of education, as in the information deficit model, or as holders of valuable knowledge). In these situations, the scope for the inclusion of lay expertise is often limited and only possible if this expertise can be translated – or scientised – so that it fits with dominant scientific understandings. This means that, in practice, participatory forms of knowledge production run the risk of reproducing the very same unequal power relations that they intended to overcome. The earlier cited literature on indigenous co-management offers a number of examples of this (Nadasdy, 2003; Fernandez-Gimenez et al., 2006). Preventing the appropriation or undue scientisation and ensuring the equal contribution of lay expertise remains a considerable challenge for participatory knowledge production. The case study of Loweswater (Case I: The Loweswater Care Project) offers an interesting example of how this challenge can be addressed. Here, a collaboration emerged where scientists and lay people generated knowledge that was self-consciously provisional, experimental, collective, and doubted, with the objective to jointly work out what 'the pollution problem' actually was. The case illustrates a relationship between science and lay expertise that respects differences and diversity between forms of knowledge without essentialising those differences, and that brings them into productive dialogue without attempting to integrate them in a formal sense (see Chapter 7).

8.6 Conclusion

In this chapter, we have used the term 'lay expertise' to denote what is often called traditional, local, indigenous, or lay knowledge. Using examples from conservation, biodiversity, and environment, this chapter has developed the following arguments:

• Like science, lay expertise has its advocates that argue for the unique qualities and values of this form of knowledge. However, by treating science and lay expertise as static categories, with universal boundaries between them, they risk ignoring the historical interdependencies and interactions between science and lay expertise.
• To move beyond these rather sterile dichotomies, it is helpful to conceptualise all forms of knowledge as situated practice. This implies an understanding of the differences between forms of knowledge as historically grown and human made, which means that these differences are not set in stone and that different knowledge systems are not in principle irreconcilable.
• Collaboration between science and lay expertise can have beneficial outcomes, but only if common pitfalls are prevented which often result in the appropriation of or limited inclusion of lay expertise. Specifically, the almost self-evident power that is associated with science has to be questioned to ensure sufficient scope for the contribution of lay expertise on equal terms.

These conclusions pave the way for Chapters 9 and 10, which deal more explicitly with the question of how the relation between science and non-science (including citizens, societal groups, and policy) can be organised to ensure the legitimacy and accountability of science in society.

References

Acheson, J. M. (2006). Institutional Failure in Resource Management. *Annual Review of Anthropology*, 35(1), 117–134.

Agrawal, A. (1995). Dismantling the Divide Between Indigenous and Scientific Knowledge. *Development and Change*, 26(3), 413–439.

Alvard, M. S. (1993). Testing the 'Ecologically Noble Savage' Hypothesis: Interspecific Prey Choice by Piro Hunters of Amazonian Peru. *Human Ecology*, 21(4), 355–387.

Berkes, F. (1999). *Sacred Ecology: Traditional Ecological Knowledge and Resource Management*. Philadelphia: Taylor and Francis.

Berkes, F., Colding, J., and Folke, C. (2000). Rediscovery of Traditional Ecological Knowledge as Adaptive Management. *Ecological Applications*, 10(5), 1251–1262.

Bonney, R., Cooper, C. B., Dickinson, J., et al. (2009). Citizen Science: A Developing Tool for Expanding Science, Knowledge, and Scientific Literacy. *Bioscience*, 59, 977–984.

Brown, T. (2014). Chipko Legacies: Sustaining an Ecological Ethic in the Context of Agrarian Change. *Asian Studies Review*, 38(4), 639–657.

Bruyninckx, J. (2015). Trading Twitter: Amateur Recorders and Economies of Scientific Exchange at the Cornell Library of Natural Sounds. *Social Studies of Science*, 45(3), 344–370.

Bryan, J. (2011). Walking the Line: Participatory Mapping, Indigenous Rights, and Neoliberalism. *Geoforum*, 42, 40–50.

Chilvers, J., and Kearnes, M., eds. (2015). *Remaking Participation: Science, Environment and Emergent Publics*. New York: Routledge.

Collins, H., and Evans, R. (2008). *Rethinking Expertise*. Chicago: The University of Chicago Press.

Downie, B. (2015). *Conservation Influences on Livelihood Decision-making: A Case Study from Saadani National Park, Tanzania*. Unpublished dissertation, Department of Geography University of Victoria.

Edgerton, R. B. (1992). *Sick Societies: Challenging the Myth of Primitive Harmony*. New York: The Free Press.

Ellingson, T. (2001). *The Myth of the Noble Savage*. London: University of California Press.

Ellis, R., and Waterton, C. (2004). Environmental Citizenship in the Making: The Participation of Volunteer Naturalists in UK Biological Recording and Biodiversity Policy. *Science and Public Policy*, 31(2), 95–105.

Fernandez-Gimenez, M., Huntington, H., and Frost, K. (2006). Integration or Co-optation? Traditional Knowledge and Science in the Alaska Beluga Whale Committee. *Environmental Conservation*, 33(4), 306–315.

Funtowicz, S., and Ravetz, J. (1993). Science for the Post-Normal Age. *Futures*, 25(7), 739–755.

Gadgil, M., Berkes, F., and Folke, C. (1993). Indigenous Knowledge for Biodiversity Conservation. *Ambio*, 22(2/3), 151–156.

Gadgil, M., Seshagiri Rao, P. R., Utkarsh, G., Pramod, P., and Chhatre, A. (2000). New Meanings for Old Knowledge: The People's Biodiversity Registers Program. *Ecological Applications*, 10(5), 1307–1317.

Gaukroger, S. (2006). *The Emergence of a Scientific Culture*. Oxford: Oxford University Press.

Gibbons, M. (1999). Science's New Social Contract with Society. *Nature*, 402, c81–c84.

Gibbons, M., Limoges, C., Nowotny, H., Schwartzman, S., Scott, P., and Trow, M. (1994). *The New Production of Knowledge: The Dynamics of Science and Research in Contemporary Societies*. London: SAGE Publications.

Goldman, M. (2007). Tracking Wildebeest, Locating Knowledge: Maasai and Conservation Biology Understandings of Wildebeest Behavior in Northern Tanzania. *Environment and Planning D: Society and Space*, 25(2), 307–331.

Gura, T. (2013). Citizen Science: Amateur Experts. *Nature*, 496(7444), 259–261.

Halffman, W. (2005). Science–Policy Boundaries: National Styles? *Science and Public Policy*, 32(6), 457–467.

Haklay, M. (2013). Citizen Science and Volunteered Geographic Information: Overview and Typology of Participation. In D. Sui, S. Elwood, and M. Goodchild, eds. *Crowdsourcing Geographic Knowledge: Volunteered Geographic Information (VGI) in Theory and Practice* (pp. 105–122). Heidelberg: Springer.

Haraway, D. (1988). Situated Knowledges: The Science Question in Feminism and the Privilege of Partial Perspective. *Feminist Studies*, 14(3), 575–599.

Herrera, P. M., Davies, J., and Baena, P. M. (2014). *The Governance of Rangelands: Collective Action for Sustainable Pastoralism*. New York: Routledge.

Irwin, A. (1995). *Citizen Science, a Study of People, Expertise and Sustainable Development*. New York: Routledge.

Jasanoff, S. (2005). *Designs on Nature, Science and Democracy in Europe and the United States*. Princeton: Princeton University Press.

Klein, J. T., Grossenbacher-Mansuy, W., Haberli, R., Bill, A., Scholz, R. W., and Welti, M. (2001). *Transdisciplinarity: Joint Problem Solving among Science, Technology, and Society*. Basel: Birkhauser Verlag.

Kloppenburg, J. (2010). Impeding Dispossession, Enabling Repossession: Biological Open Source and the Recovery of Seed Sovereignty. *Journal of Agrarian Change*, 10 (3), 367–388.

Lane, S. N., Odoni, N., Landström, C., Whatmore, S. J., Ward, N., and Bradley, S. (2011). Doing Flood Risk Science Differently: An Experiment in Radical Scientific Method. *Transactions of the Institute of British Geographers*, 36(1), 15–36.

Latour, B. (1993). *We Have Never Been Modern*. Boston: Harvard University Press.

Lawrence, A. (2010). *Taking Stock of Nature: Participatory Biodiversity Assessment for Policy Planning and Practice*. Cambridge: Cambridge University Press.

Lawrence, A., and Turnhout, E. (2010). Personal Meaning in the Public Sphere: The Standardisation and Rationalisation of Biodiversity Data in the UK and the Netherlands. *Journal of Rural Studies*, 26(4), 353–360.

Milton, K. (2002). *Loving Nature: Towards an Ecology of Emotion*. London and New York: Routledge.

Moore, K. (2009). *Disrupting Science: Social Movements, American Scientists, and the Politics of the Military*. Princeton: Princeton University Press, 1945–1975.

Nadasdy, P. (1999). The Politics of TEK: Power and the 'Integration' of Knowledge. *Arctic Anthropology* 36, 1–18.

Nadasdy, P. (2003). Reevaluating the Co-Management Success Story. *Arctic*, 56(4), 367–380.

Nandy, A., ed. (1988). *Science, Hegemony and Violence: A Requiem for Modernity*. United Nations University.

Neves-Graca, K. (2004). Revisiting the Tragedy of the Commons: Ecological Dilemmas of Whale Watching in the Azores. *Human Organization*, 63(3), 289–300.

Neves-Graca, K. (2006). Politics of Environmentalism and Ecological Knowledge at the Intersection of Local and Global Processes. *Journal of Ecological Anthropology*, 10, 19–32.

Neves, K. (2010). Cashing in on Cetourism: A Critical Ecological Engagement with Dominant E-NGO Discourses on Whaling, Cetacean Conservation, and Whale Watching. *Antipode*, 42(3), 719–741.

Ottinger, G. (2010). Buckets of Resistance: Standards and the Effectiveness of Citizen Science. *Science, Technology & Human Values* 2010, 35, 244–270.

Pahl-Wostl, C. (2015). *Water Governance in the Face of Global Change: From Understanding to Transformation*. Cham: Springer.

Pellizzoni, L., 2001. The Myth of the Best Argument: Power, Deliberation and Reason. *British Journal of Sociology*, 52(1), 59–86.

Ponting, J., and O'Brien, D. (2014). Liberalizing Nirvana: An Analysis of the Consequences of Common Pool Resource Deregulation for the Sustainability of Fiji's Surf Tourism Industry. *Journal of Sustainable Tourism*, 22(3), 384–402.

Raffles, H. (2002). Intimate Knowledge. *International Social Science Journal*, *54*(173), 325–335.

Rangan, H., ed. (1996). From Chipko to Uttaranchal. In R. Peet and M. Watts, eds., *Liberation Ecologies: Environment, Development and Social Movement*. London: Routledge, 205–226.

Redford, K. H. (1991). The Ecologically Noble Savage. *Cultural Survival Quarterly*, 15 (1), 46.

Ruttan, L. M. (1998). Closing the Commons: Cooperation for Gain or Restraint? *Human Ecology*, 26(1), 43–66.

Shirk, J., Ballard, H., Wilderman, C., et al. (2012). Public Participation in Scientific Research: A Framework for Deliberate Design. *Ecology and Society*, 17(2), 29.

Shiva, V., and Bandyopadhyay, J. (1986). The Evolution, Structure, and Impact of the Chipko Movement. *Mountain Research and Development*, 6(2), 133.

Shiva, V. (1993). *Monocultures of the Mind: Perspectives on Biodiversity and Biotechnology*. London: Zed Books.

Shiva, V. (1997). *Biopiracy the Plunder of Nature and Knowledge*. Cambridge: South End Press

Star, L. S., and Griesemer, J. R. (1989). Institutional Ecology, 'Translations' and Boundary Objects: Amateurs and Professionals in Berkeley's Museum of Vertebrate Zoology. *Social Studies of Science*, 19(3), 387–420.

Stirling, A. (2006). Analysis, Participation and Power: Justification and Closure in Participatory Multi-Criteria Analysis. *Land Use Policy*, 23, 95–107.

Stirling, A. (2008). 'Opening Up' and 'Closing Down': Power, Participation, and Pluralism in the Social Appraisal of Technology. *Science, Technology & Human Values*, 33, 262–294.

Toogood, M. (2013). Engaging Publics: Biodiversity Data Collection and the Geographies of Citizen Science. *Geography Compass*, 7(9), 611–621.

Tsouvalis, J., and Waterton, C. (2012). Building 'Participation' Upon Critique: The Loweswater Care Project, Cumbria, UK. *Environmental Modelling & Software*, 36, 111–121.

Turnbull, D. (1997). Reframing Science and Other Local Knowledge Traditions. *Futures*, 29, 551–562.

Turnbull, D. (2009). Futures for Indigenous Knowledges. *Futures*, 41, 1–5.

Turnhout, E., Lawrence, A., and Turnhout, S. (2016). Citizen Science Networks in Natural History and the Collective Validation of Biodiversity Data. *Conservation Biology*, 30(3), 532–539.

Turnhout, E., Stuiver, M., Klostermann, J., Harms, B., and Leeuwis, C. (2013). New Roles of Science in Society: Different Repertoires of Knowledge Brokering. *Science and Public Policy*, 40(3), 354–365.

Van der Ploeg, J. D. (2002). Potatoes and Knowledge. In M. Hobart, ed., *An Anthropological Critique of Development: The Growth of Ignorance* (pp. 209–224). New York: Routledge.

Wall, D. (2014). *The Commons in History: Culture, Conflict, and Ecology.* Harvard: The MIT Press.

Whatmore, S. J., and Landström, C. (2011). Flood Apprentices: An Exercise in Making Things Public. *Economy and Society,* 40(4), 582–610.

Wynne, B. (1996). May the Sheep Safely Graze? A Reflexive View of the Expert–Lay Knowledge Divide. In S. Lash, B. Szerszynski, and B. Wynne, eds., *Risk, Environment and Modernity, Towards a New Ecology* (pp. 44–83). London: SAGE Publications.

Yang, L. (2015). Local Knowledge, Science, and Institutional Change: The Case of Desertification Control in Northern China. *Environmental Management,* 55(3), 616–633.

Case H Lay Expertise and Botanical Science
A Case of Dynamic Interdependencies in Biodiversity Conservation
KATJA NEVES

This case describes several programmes at botanical gardens throughout the world that bring together scientific and lay knowledge of biodiversity. Lay experts, such as gardeners and volunteers, operate in complex cooperation patterns with professional scientists, such as ecologists and botanists. Neves shows that it is neither easy nor fruitful to try to draw a sharp line between these communities, and that complex forms of cooperation have developed, even though some knowledge hierarchies may still be present. These examples of the interaction between lay and professional expertise show how productive collaboration is possible, with important contributions to areas such as biodiversity conservation, horticulture, and ecological restoration.

Introduction

As seen in Chapter 8, despite the socio-cultural groundedness of all forms of scientific and non-scientific knowledge, the literature on lay ecological knowledge continues to operate by contrasting lay expertise with scientific knowledge. According to this literature, this made sense because it enabled these authors to highlight what they saw as crucial differences between these knowledges. The distinction highlighted the different rationalities in which these knowledge systems are embedded, and it offered a way for advocates of indigenous and local communities to unveil the uniqueness and value of these knowledge practices and, in doing so, give the holders of this knowledge a voice (Battiste, 2000).

As the field of lay ecological knowledge studies developed, however, it became increasingly evident that it was crucial to develop strategies for these distinct but complementary knowledges to communicate and interact productively (e.g. Neves-Graca, 2004). From this perspective, the recognition of the validity of multiple knowledges could offer scope for greater democratic participation in knowledge production processes, which could then ultimately lead to solutions that represent the different views and interests that are at stake. This view has influenced the institutionalisation of organisations such as the Intergovernmental Platform for Biodiversity and Ecosystem Services (IPBES) and Botanical Gardens Conservation International (BGCI), which have developed structures and communication mechanisms that open up space to explore the complementarity of lay and expert/scientific knowledge. To support these arguments, this chapter discusses the case of botanical gardens as an illustration of the entanglements that have existed historically between lay expertise and scientific knowledge.[1]

The Historical Entwinement of Science and Lay Expertise in Botanical Gardens

To date, most scholarly research on botanical gardens has focused on the historical roles botanical gardens played in training medical doctors in the use of medicinal plants, in developing botany as a field of scientific endeavour, and in empire building (e.g. Grove, 1995; Baber, 1996, 2001; Brockway, 1979; Griffiths and Robin, 1997). Scholars working in this context have highlighted the role of these institutions in developing universal systems of plant classification that rendered 'New World' flora commensurable with European understandings, valuations, and uses of plants at the detriment of socio-ecologically embedded knowledges and practices (Brockway, 1979; Cascoigne, 1998; Neves, 2009, 2012, 2019). However, historical research also demonstrates that botanical gardens have depended heavily on lay expertise for centuries. In fact, productive interactions between lay expertise and scientific knowledge practices span centuries. One must note, however, that more often than not these interactions were marked by exploitative, unequal, and often violent relations between holders of lay expertise and institutions of scientific knowledge.

The case of cinchona, from which quinine is produced, is a case in point in this context as the extraction of local lay expertise by botanical gardens, and even of cinchona tree seeds, was done against the laws of the countries from which they were extracted (e.g. the case of Brazil). As mentioned in Chapter 8, such cases are often called 'biopiracy'. The extraction of cinchona from tree bark for the production of quinine was a complicated process that was totally

---

[1] The materials presented and discussed herein are based on my forthcoming book (Neves, 2019), which offers in-depth analysis of the role that botanical gardens played in the rise of modernity, the nation state, empire, and capitalism, and, more recently, in the emergence of transnational systems of plant biodiversity conservation governance.

unknown to Europeans during the initial phases of colonial expansion (Rocco, 2003). Nevertheless, as quinine proved to be extremely effective in combating malaria – and, as such, was crucial to the expansion of European empires over tropical areas – Kew Royal Botanic Gardens UK sent plant hunters to South America to learn from local lay experts not only the secrets behind producing quinine from cinchona bark, but also the secrets of planting and growing these trees. While it is true that British botanists subsequently improved cinchona trees so as to extract higher concentrations of quinine and medical scientists refined the knowledge of quinine application, these developments were built on the solid foundations of local lay expertise and practice. Similar interactions between lay expertise and scientific knowledge ground the colonial histories of rubber, tea, coffee, and cotton, to name but a few.

More recently, scholars have turned their attention to the association of botanical gardens with resistance to colonialism and with the emergence of more equal forms of 'collaboration' between lay expertise and scientific knowledge (Baber, 1996, 2001; Connan, 2008). Institutions such as Brazil's Rio de Janeiro Botanical Garden (Gaspar and Barata, 2008), Canada's VanDusen Botanical Garden (Khaledi, 2008), and Australia's Royal Botanic Gardens Victoria at Cranbourne (Whillougby, 2014), which hold centuries-old histories in the (re)valuation of native/aboriginal plant knowledge practices, are currently working to establish international networks that disrupt the old power dynamics associated with the expansion of scientific power as well as with exclusionary nation/empire-building processes. The necessity and importance of these collaborations has gained wider traction within the context of the Global Strategy for Plant Conservation (GSPC). As leaders in the development and implementation of the GSPC (Neves, 2012, 2014a, 2014b), botanical gardens are currently developing initiatives to foster citizen participation and the inclusion of lay expertise in their work.

Harnessing Lay Expertise: Recent Examples from Botanical Garden Biodiversity Conservation

An important reason why botanical gardens offer such a productive venue for collaboration between scientific knowledge and lay expertise is their open and nuanced understandings of what conservation is and who can contribute (Neves, 2009, 2012, 2019). This pertains both to the cultivation of protected species inside their walls (ex situ conservation), and to the conservation of plants within the ecosystems that they are part of (in situ conservation)

(Maunder, 1994, 2008). While dominant approaches to conservation focus on purportedly pristine natures and/or wilderness and are predominantly informed by natural science experts in biology and ecology, botanical gardens have always been positioned at the intersection of nature and culture and have blurred the boundaries between them. For example, a growing number of botanical gardens around the world see the private gardens of urban dwellers and citizens in general as sites that can make important contributions to biodiversity.

One such example is the Toronto Botanical Gardens (TBG) in Canada, which has developed comprehensive curricula to teach amateur and master gardeners how to engage in gardening practices that simultaneously achieve leisure, aesthetic, and conservation goals. These strategies range from motivating lay experts to plant flowers, trees, and other plant materials that help provide habitats for local ecologically important species such as butterflies, birds, insects, invertebrates, fungi, and soil bacteria to mixing food crops with display plants so as to minimise the ecological footprint of urban dwellers. TBG also nurtures engagement with environmentally conscious practices such as making climate-ecologically adequate choices pertaining to gardening, and/or choosing endemic flora in need of restoration. Teaching people how to choose locally adapted grasses to reduce watering and fertilisation loads further exemplifies this approach. In turn, the private gardens of urban citizens have the added potential of providing much-needed biodiversity corridors for threatened migratory species, such as the charismatic monarch butterfly. Even though TBG's approach to amateur horticultural/leisure gardening spearheads a newly unfolding trend, it is far from unique or rare. An ever-expanding number of botanical gardens around the world have developed, or are developing, similar approaches.[2]

In addition to changing understandings of who may constitute effective agents of biodiversity conservation beyond the narrower scope of scientists, bureaucrats, and expert ecosystem managers, profound transformations are occurring in the recognition of the validity and usefulness of non-scientific knowledges, including lay expertise. Recognition of the role of knowledge complementarity within the botanical-garden world is exemplified by a growing number of botanical gardens that rely on the general public to collect critically important scientific data and/or participate in scientific interventions. For example, Espace Pour La Vie Montreal Botanical Gardens developed a programme whereby the general public plays a crucial role in tagging

---

[2] See www.missouribotanicalgarden.org/things-to-do/events/2015-bgci-congress/program.aspx for examples.

monarch butterflies so that their migratory paths can be better understood. Citizens in southern Quebec and Ontario also participate in the collection, identification, and location of tagged butterflies to these ends.

Another example is provided by the New York Botanical Gardens, which developed a programme that relies on the recruitment of regular citizens to collect data on eight species of native trees that are being studied in relation to the impacts that climate change has on US native forests. Citizens participating in this project not only learn about scientific methods of data collection, but also about the results of these scientific efforts and, in so doing, about the effects of climate change on their local environments.

The above are but two examples of what is now called 'citizen science', which the Cornell Lab of Ornithology (n.d.) incisively defines as a term that

> has been used to describe a range of ideas, from a philosophy of public engagement in scientific discourse to the work of scientists driven by a social conscience.
> In North America, citizen science typically refers to research collaborations between scientists and volunteers, particularly (but not exclusively) to expand opportunities for scientific data collection and to provide access to scientific information for community members.

There are many more such collaborations in botanical gardens, many of which are long standing, but often invisible. Here, I refer specifically to the role that lay expertise plays within the often unequal structure of botanical-garden knowledge hierarchies, which to date still put scientific knowledge on top. Botanists and ecologists working at botanical gardens rely heavily on the lay expertise of the gardeners and other staff also working at these institutions. One might even argue that what is often perceived as the product of scientific labour is in fact very often the product of scientific collaboration with lay expertise. Two additional cases pertaining to contemporary botanical garden biodiversity conservation illustrate this point; one is located in the Portuguese archipelago of the Azores, the other in Lyon, France.

The Jardim Botânico do Faial in the Azores (Portugal) is a prime example of a botanical garden that is simultaneously engaged with ex situ and in situ biodiversity conservation while also pursuing an important and innovative programme of ecosystem restoration. BGCI (n.d.) describes Jardim Botânico do Faial's mission as centred on 'the management of the collection of living flora associated with botanical conservation, especially the collection of seeds of endemic species, the recovery of habitats and the sensitization to the importance of natural floristic wealth of the Azores'. The Jardim Botânico do Faial has in fact taken on a leading role in the development of ecological restoration measures within the wider ecological context of the Macaronesia, the phytogeography region comprised of the Portuguese archipelagos of the Azores, Madeira, the Spanish archipelago of the Canary Islands, and the Republic of Cape Verde, an island country situated off the coast of east Africa.

The Jardim's highly successful and important ecological project is to a great degree the result of a productive collaboration between scientists and lay experts. More precisely, it is the outcome of a serendipitous joining of conservation efforts between scientists working within the botanical garden and a small group of local 'forest and nature rangers'.[3] Working within the scope of the regional ministry of the environment, the local forest and nature rangers were responsible for a series of tasks that effectively left them imbued with a strong sense of environmental 'stewardship' (as a key participant in my research put it). Although their posts were initially created in the late 1980s to invigilate citizen conduct in relation to conservation norms, it soon also encompassed the roles of educating populations, managing endemic ecological niches by keeping them clean of invasive plants and other destructive dynamics, and reproducing endemic plants for subsequent restoration.

The latter – the reproduction of endemic plants for subsequent restoration – has presented enormous challenges to many botanical gardens around the world. Endemic plants are often non-domesticated species that are therefore not as easily reproduced and or re-locatable. While botanical garden scientists may know a great deal about the botany of these species, or even about their unique ecological dynamics, they are often at a loss in terms of how to actually reproduce (cultivate) them on the large scale that most restoration programmes demand. This was the case for the Jardim Botânico do Faial, which was founded with the core missions of education, public dissemination, and, above all, endemic plant research and restoration.

Quite fortuitously, in embracing the task of reproducing endemic plants for subsequent conservation the aforementioned forest and nature rangers chose to undergo a painstakingly laborious process of trial and error that extended over a period of more than 20 years. Even when their mandate was reformulated to much narrower terms in the early late 1990s, and they were no longer required to engage in plant restoration, some continued to care for the informal experimental gardens they had planted in areas surrounding the very ecosystems they were hired to protect. Through time these informal projects – often pursued during a leading forest nature ranger's spare time – faded into obscurity and were almost completely abandoned.

---

[3] The actual local term for this category is *Guarda e Vigilante da Natureza*. It does not easily translate into English. Helder Fraga, a local *Guarda e Vigilante da Natureza* and a key participant in my research project, described it as a kind of 'environmental guard that conciliates policing with educational and gardening mandates'.

They nevertheless came to play a central role when the Jardim Botânico do Faial was taken over by new leadership keen on fully implementing the garden's mission of pursuing research on endemic plants and of engaging in endemic plant restoration. Comprised of a relatively junior directorate with degrees in environmental sciences, the gardens' new leadership quickly realised the importance of the lay expertise that the forest and nature rangers had developed over the years. Although collaboration between scientists and these lay experts often entailed friction around issues such as clashing understandings of knowledge authorship or the onus of actual endemic ecosystem management, the two groups have been able to work productively together to develop a comprehensive programme of ecosystem restoration in the archipelago. Even if at times punctuated by reluctance, the Jardim's scientists have recognised the priceless contribution of the practical knowledge that the forest rangers developed over the years on how to reproduce and relocate endemic plants that had never been domesticated before and which resisted cultivation. The forest and nature rangers, in turn, came to recognise the importance of scientific knowledge for the wider aspects of ecological dynamics of restoration, its relation to other environmental processes, and its place within the broader context of the Macaronesia phytogeography region. Dispute and contention may remain regarding how best to approach different restoration challenges, but the different constituencies have reached agreement on how to negotiate differences of opinion with an understanding that they are within the framework of a botanical garden structure that values their contributions hierarchically.

The Jardin Botanique de Lyon, in France, provides the second example of the current recognition of the growing importance of the synergies that can take place between lay expertise and scientific knowledge.[4] As with many botanical gardens, plant biodiversity conservation has become one of the Jardin Botanique de Lyon's core missions. In this context the Jardin hosts thousands of plant species, collected from around the world over a period of more than three centuries. The current preservation of this plant material is a manifestation of the Jardin's deep commitment to ex situ conservation and the conservation of species whose habitats may be threatened or destroyed.

While this form of plant conservation is characteristic of many botanical gardens around the world, the Jardin Botanique is also part of a national

---

[4] Data concerning the *Jardin Botanique de Lyon* and its relation to *Conservatoire des Collections Végétales Spécialisées* was obtained in the form of in-depth video recorded interviewing with the *Jardin*'s former director Frederic Pautz in the summer of 2014, who is also the president of this organisation's *Comité des Collections* (see www.ccvs-france.org). Additional information was also obtained from *Jardin Botanique de Lyon*'s website, including the related documentaries it offers.

network that focuses specifically on the identification, preservation, and restoration of France's botanical and horticultural patrimony. This network is overseen by the Conservatoire des Collections Végétales Spécialisées (CCVS), an umbrella organisation that coordinates and harmonises the public and private conservation of this patrimony at a national level. Given the focus of the Jardin Botanique and the CCVS on the preservation of cultivars (i.e. plants species that have developed by means of human-directed selective breeding across several centuries) these institutions are effectively also pursuing in situ conservation of what can adequately be described as France's bio-cultural diversity. If one takes into account that the evolution of cultivars is the cumulative effect of the informal practices of gardeners and farmers across decades, even centuries, and that many of these cultivars are still grown today, it becomes evident that not only does this project further blur the distinction between in situ and ex situ conservation, but also that between lay expertise and scientific knowledge.

In effect, the Jardin Botanique is one of many French organisations working within the scope of the CCVS to collaborate in the coordination of lay expertise and scientific knowledge when it comes to the preservation of France's bio-cultural diversity. A key dimension of this initiative has been the creation of a database with practical information about the specific conditions under which specific cultivars are grown and thrive, about how best to reproduce these species, and about how to engage in their restoration. This kind of knowledge has mostly been collected by lay experts (e.g. gardeners and farmers), and is subsequently complemented by scientific knowledge produced by experts working in botanical garden departments; the latter knowledge is then also shared with lay experts. In these circles, it is widely recognised that the preservation of France's bio-cultural patrimony would be considerably hampered by the absence of such collaborations.

Concluding Remarks

This case, and Chapter 8 more broadly, have offered a number of examples to illustrate that scientific and lay expertise are not irreconcilably different and are not demarcated by clear boundaries. Rather, the emphasis that is often put on the differences between these two knowledges is to a great extent the result of historical–political contingency whereby the similarities were overshadowed for specific socio-cultural and political reasons. The case of botanical gardens not only demonstrates the historical co-dependencies of lay and scientific expertise, but also shows that the reasons for demarcating science from lay expertise and attributing greater authority to science were often related to the expansion of empire and exploitation. The recent shift in the botanical-gardens world towards biodiversity conservation and participatory approaches has brought to light the crucial importance of exploring synergies between different knowledge practices. While these relations can at times be difficult and are often riddled with unequal distribution of power, they are becoming increasingly central in this era of environmental concern.

References

Baber, Z. (2001) Colonizing Nature: Scientific Knowledge, Colonial Power, and the Incorporation of India into the Modern World System. *British Journal of Sociology*, 21(1), 37–58.

Baber, Z. (1996). *The Science of Empire: Scientific Knowledge, Civilization, and Colonial Rule in India*. Albany: State University of New York Press.

Battiste, M. A., ed., (2000). *Reclaiming Indigenous Voice and Vision*. Vancouver: Univeristy of British Columbia Press.

BGCI (n.d.). *About the Jardim Botânico do Faial*. Botanic Gardens Conservation International. www.bgci.org/garden.php?id=1465

Brockway, L. H. (1979). *Science and Colonial Expansion: The Role of the British Royal Botanic Gardens*. New York: Academic Press.

Cascoigne, J. (1998). *Science in the Service of Empire: Joseph Banks, the British State and the Use of Science in the Age of Revolution*. Cambridge: Cambridge University Press.

Conan, M., ed. (2008). *Gardens and Imagination: Cultural History and Agency*. Cambridge: Harvard University Press.

Gaspar, C. B., and Barata, C. E. (2008). *De engenho a jardim: memorias historica do Jardim Botanico*. Rio de Janeiro: Capivara Editora.

Griffiths, T., and Robin, L., eds. (1997). *Ecology and Empire: Environmental History of Settler Societies*. Seattle: University of Washington Press.

Grove, R. (1995). *Green Imperialism: Colonial Expansion, Tropical Island Edens, and the Origins of Environmentalism, 1600–1860*. Cambridge: Cambridge University Press.

Khaledi, B. (2008). The Colonial Present: Botanical Gardens as Sites of Nationalism, Environmentalism and Aboriginality in British Columbia. Unpublished MA Thesis, Department of Sociology and Anthropology, Simon Fraser University.

Maunder, M. (1994). Botanic Gardens: Future Challenges and Responsibilities. *Biodiversity and Conservation*, 3(2), 97–103.

Maunder, M. (2008). Beyond the Greenhouse. *Nature*, 455(7213), 596–597.

Neves-Graca, K. (2004). Revisiting the Tragedy of the Commons: Ecological Dilemmas of Whale Watching in the Azores. *Human Organization*, 63(3), 289–300.

Neves, K. (2009). Urban Botanical Gardens and the Aesthetics of Ecological Learning: A Theoretical Discussion and Preliminary Insights from Montreal's Botanical Garden. *Anthropologica*, 51(1), 145–157.

Neves, K. (2012). The New Roles of Botanical Gardens in Biodiversity Conservation. In P. Robbins, ed., *Encyclopedia of Global Climate Change*. SAGE Publications Online Encyclopedia.

Neves, K. (2014a). A grande saga wollemi: entre a preservação de genoma e a conservação consumista. *Memorias Special Journal Issue on Environmental Anthropology*, 105–118.

Neves, K. (2014b). Horta: Notes on an Invaluable Legacy. *Mundo Acoreano* (special issue).

Neves, K. (2019). *Post-normal Conservation: Botanic Gardens and the Reordering of Biodiversity Conservation*. New York: State University of New York Press.

Rocco, F. (2003). *Quinine: Malaria and the Quest for a Cure that Changed the World*. New York: Harper Perennial.

The Cornell Lab of Ornithology. (n.d.). *Defining Citizen Science*. www.birds.cornell.edu/citscitoolkit/about/definition

Willoughby, S. (2014). *Borders and Boundaries in Botanic Gardens Conservation*. Paper presented at Leaders in Conservation Conference, Montreal Canada October 2014.

Case I The Loweswater Care Project

CLAIRE WATERTON

Environmental protection is not just a technical issue, but also a social concern to which local inhabitants can make an important contribution. This case documents a participatory approach, involving natural scientists, social scientists, and other local stakeholders, to develop a joint understanding of an environmental problem in Loweswater, one of the lakes of the English Lake District. The approach was specifically developed to prevent some of the problems of participation discussed in Section 8.5, and offers important lessons about how to ensure meaningful collaboration and how to co-produce environmental knowledge.

Introduction

This case study concerns the making of a forum initially called a 'knowledge collective' at Loweswater, a small village in the north-west of the Lake District in Cumbria, in the north of England.[1] This forum was established to address the problem of pollution in a local lake (see Figure I1), but it was established with the strong underpinning that all forms of expertise brought to the forum were to be equally welcomed, valued, doubted, and debated. As Chapter 8 has argued, the categories of scientific knowledge and lay expertise tend to be unhelpfully static, and this is particularly the case if lay and professional people are organising together to put new questions to specific problems in which they are mutually interested. As several chapters in this volume have attested (see, for example, Chapter 8), such situations are

---

[1] The research on which this chapter is based was carried out by an interdisciplinary team of six researchers (Claire Waterton, Department of Sociology – Lancaster University; Stephen Maberly, Centre for Ecology and Hydrology – Lancaster Environment Centre; Judith Tsouvalis, School of Sociology and Social Policy – University of Nottingham; Nigel Watson, Lancaster Environment Centre; Ian Winfield, Centre for Ecology and Hydrology – Lancaster Environment Centre; and Lisa Norton, Centre for Ecology and Hydrology – Lancaster Environment Centre), and one 'community researcher', Ken Bell, a farmer based in Loweswater.

Figure I1 Loweswater in the English Lake District (photo by Lisa Norton).

increasingly common in the current 'age of participation' where different actors, embodying different kinds of expertise, are being enrolled, and enrolling others, into processes of problem solving and decision making (see also Chilvers and Kearnes, 2015). Thus, we might say that this case study – concerning the creation of a knowledge collective around a polluted lake from 2007 onwards – partly added to a plethora of participatory initiatives that were sprouting up in the early 2000s around local environmental problems. But, as I will discuss, it also in part represented an attempt to make the politics of knowledge-making within that initiative more explicit and to deliberately iron out hierarchies of expertise.

At Loweswater in the early 2000s, however, a more material issue was also at stake. The creation of a knowledge collective was felt to be a necessary response to the ecological changes evidently underway in the small lake, Loweswater, situated within the village of Loweswater, in what is known to be a 'tranquil and unspoilt corner of the Lake District'.[2] Since the early 2000s this lake, a recognised beauty spot, had been hosting increasingly high populations of cyanobacteria. These microscopic organisms were manifesting on the surface of the lake as an 'algal bloom' (a pea-green slime covering the surface of the water), which could be toxic to large mammals (including cows, dogs, and on occasion, humans). Also, since their presence was contributing to the failure of Loweswater's freshwater to meet the criterion of 'Good Ecological

[2] www.lakedistrict.gov.uk/

Status' required under the European Water Framework Directive,[3] these bacteria had been the focus of much discussion, monitoring and research.

But this research was contested. By 2007, when the Loweswater knowledge collective was being considered as a possibility, land and lake ecologists from the Centre for Ecology and Hydrology at Lancaster University had been working at Loweswater for several years trying to determine which practices and processes on the land around the lake were supporting the high populations of potentially toxic cyanobacteria within it. The Environment Agency, responsible for the implementation of the recent European Water Framework Directive in England and Wales, had also been collecting data since the early 2000s, monitoring the biological and chemical composition of the lake water at regular intervals. However, it was clear at that time that some local people felt a deep cynicism about this ongoing scientific research. In response to a short survey sent out in 2007 to local residents about the possibility of creating a local forum for discussion of the pollution issues in the lake, one resident wrote: 'In three years' time, one of three things will happen to the algae, the consequences of which are the same: 1) Situation improve – requires further monitoring; 2. Situation deteriorates – requires further monitoring; 3) Situation stays same – requires further monitoring.'[4] At the same time, farmers were antagonised by the actions of the Environment Agency, who not only monitored the lake water, but also sent letters to those farmers considered to be in breach of farm-waste regulations, warning them of penalties and prosecution unless they changed their waste-management practices.

What we can see here, then, is a situation where scientific expertise was needed locally (to determine trends of cyanobacteria and the nutrients supporting them), but was also regarded as alien – and alienating – to local experience and local practices. A tension existed that had the effect of separating ideas of 'scientific' and 'lay' knowledge. This tension was not insurmountable, however, and it did not mean that local residents were not open to the idea of working with scientists – in this case, ecologists and Environment Agency staff. Given the opportunity to do so from 2007 onwards through a new publicly funded project,[5]

---

[3] European Water Framework Directive (WFD) was issued by the European Parliament and European Council (2000/60/EC) in 2000. This directive imposes legally binding demands on all EU member states to achieve 'good ecological status' (an internationally comparable ecological state) in European freshwater bodies. Partly because of the presence of cyanobacteria, Loweswater looked likely to fall below this European water-quality standard.

[4] Questionnaire sent out March 2007. The response was given in answer to Question 3 'Do you have any comments about the [three year, 2007–2010, RELU] project overall?' See next footnote for project details.

[5] The research 'Understanding and Acting in Loweswater: a community approach to catchment management' was supported by the Rural Economy and Land Use Programme from July 2007 to December 2010 (RES-229–25-0008).

local residents, farmers, environmental agency representatives, local business owners, and natural and social scientists began to work together in a 'new collective' that aimed to grapple with the problem of Loweswater's deteriorating water quality. Below, I describe how this collective worked in such a way as to allow for a collective and gradual reappraisal of 'the algae problem'. The activities undertaken by the collective included many inquiries, investigations, and practical activities, all of which contributed to the dissolving and re-assembling of the boundaries between scientific and lay expertise.

The Loweswater Care Project and Its Mode of Working

The knowledge collective at Loweswater was the product of a collaboration between farmers, local residents, environmental agencies, and natural and social scientists from 2007 onwards. In 2008 the collective was renamed the Loweswater Care Project (hereafter LCP) by its participants. As suggested earlier, a feature of the LCP was that all participants were assumed to be authorities: that is, all forms of expertise were to be equally valued. However, as we shall see, all forms and representations of knowledge were also to be doubted and debated. The confidence to establish such principles or 'rules of engagement' came from theoretical work in the domain of Science and Technology Studies (STS). In the mid-2000s many STS authors were writing challenging texts asking for new political configurations (or deliberately styled 'fora') that would do 'knowing' and 'acting' differently (e.g. Mol, 2002; Latour, 2004; Stengers, 2005). The LCP was, in effect, an experiment to create such a new forum.

From the early meetings of the LCP, participants adopted a set of principles derived from the publication of the science studies scholar Bruno Latour (2004). The basic goal of such principles was to prevent what Latour calls a 'foundationalist model of knowledge'. This model in many ways resembles the linear model of science–society relations discussed in Chapter 4 and holds that science is the only authoritative source of knowledge for policy and decision making. In this model, scientific knowledge holds a lot of power. Such is the power of this knowledge that it can render debate, or politics, about the matter at hand irrelevant. In other words, if science is allowed to speak in the name of some public matter, then nature, or bodies, or other matters often remain mute and uncontested (Latour, 2004).

In the *Politics of Nature* (2004), Latour argued that this situation should be turned around. This meant that nature could be seen as something by which politics could be regenerated. This could not take place through a singular form of science, as in the linear model, but it could take place through the sciences (in the plural). The proposed model is not one of science speaking truth to power, but one in which knowledge-making is about formulating proposals and testing them out (Latour, 2004). To enable this to take place in practice (i.e. for nature to regenerate politics), in Loweswater a number of simple principles were adopted. These principles, taken from Latour (2004), established that the LCP forum would be one:

- where nature is not self-evident;
- where knowledge and expertise has to be debated;
- where it is accepted that uncertainty is the main condition humans are in (rather than a condition of having knowledge);
- where what is important is the creation of connections between people and things;
- where doubt and questioning are extended to our own representations.

What can be seen in these principles is that they have the potential effect of levelling out hierarchies of expertise whilst retaining a kind of collective spirit of inquiry and ability to act and create. By suggesting that nature can never be self-evident, the forum agreed to question and query all knowledge claims, including those formally represented by science. Through these principles, the forum got used to the idea that nature cannot be represented by certain forms of scientific knowledge that might stifle debate. The forum treated knowledge claims democratically: all representations – be they lay or scientific – had to be doubted and questioned, encouraging all participants in the forum to have a say. And lastly, the forum was committed to acts of collective agency: doing things together and creating connections of many kinds between people and things.

LCP meetings were held in Loweswater village hall. Once every two months up to 40 people would gather to talk, argue, demonstrate, plan activities, report back on activities, and think about Loweswater. Participants carried out the discussions, representations, and critiques in a circle. All participants, in effect, were experts. All forms of expertise were recognised as being context-dependent, as coming *not* from nowhere but from somewhere (Haraway, 1988), and as being made in emergent co-production with other ongoing material relations (Barad, 2007).

Opening and Connecting

In practice, the LCP's bi-monthly meetings had to be structured so as to allow as much 'voice' from participants as possible, and as much listening and response as could be contained in a session lasting from 5.30 p.m. to 9.00

p.m. The meetings encouraged issues to be opened out, and connected: they supported science (with a small 's', meaning the making and testing of propositions) to be carried out through the diverse sources of expertise within the collective.

At the first formal LCP meeting, for example, a collective discussion was created through a focus on objects, the 'things' that Latour had speculated could revitalise politics (Latour, 2004). Prior to this meeting, participants had been invited (by postcard and through the parish newsletter) to bring to the meeting objects that held significance for them in terms of their relationship to the lake and its surrounding area. Many people brought objects to the forum: photographs, documents, news-cuttings, house deeds, pieces of pottery, old farm equipment, postcards, paintings – even a tiny snail that lives in the catchment, but that can only be seen under a microscope (the microscope was brought along too!). Having displayed all these things on tables in the village hall, there was plenty of time left for people to speak about their objects and the connections they wanted to make to the lake and its surroundings (see Figure I2).

Participants came to the front of the village hall to speak. Danny Leck, a local farmer who had previously worked with other Loweswater farmers to mitigate nutrient losses from land to the lake, began the work of opening out 'the algae problem'. He spoke about a photograph of his father ploughing a field with

Figure I2 Participants of the LCP view and discuss the objects that were brought to the meeting (photo by Claire Waterton).

a horse, noting the changes in farming that his family had witnessed in Loweswater, especially in terms of the intensification of beef and sheep farming in the valley (ploughing for arable crops was now infrequently seen). But he also noted that the lake has what ecologists call a long 'retention' or 'flushing' time – that is, the mean time that it takes one molecule of water to pass through the lake from the main 'inflow' to the 'outflow': 'One thing that we have found out is that Loweswater takes 180 days [to flush]'. This was compared to that of other lakes (Loweswater's retention time, for example, is thought to be six times greater than that of Grasmere, a neighbouring lake of roughly the same surface area). But Leck then made a connection to what goes into the lake, suggesting that, as a result, 'every chemical that is going into Loweswater has a bigger impact'. He spoke of the vegetation at the edge of the lake that had, in the past, been cleared regularly, but that had recently been left to grow more naturally. He aired the concern that, because of Loweswater's long retention time, vegetative matter would not get flushed through the lake quickly, suggesting that it would rot in the water, causing ecological effects including enrichment of the water.

Many people listening might possibly have thought that the effect of rotting vegetation around Loweswater could be outweighed by the effect of soluble nutrients reaching the lake from fertilised farm land (more on this below). But nevertheless, they also recognised that Leck was foregrounding an important theme: namely, that Loweswater is a water body that, by virtue of its physical constraints, is highly sensitive to all 'inputs', and therefore requires careful management. Leck suggested that the maintenance of vegetation growth, stream ('beck') walls, and drains around the lake needed to be seen as part of that management. And he recognised, as did many others who knew the local area intimately, that such maintenance had been carried out less frequently in recent years.

There are many ways in which one could analyse the set of observations that Leck made. One way, however, is to note that what he brought to the forum was not science – at least, not in the sense that it would stand for the 'facts of the matter' and require no further discussion. Indeed some ecologists present admitted that the suggestions he was making could not be supported scientifically. But rather than try and establish inscrutable 'facts', Leck allowed room for the making of propositions that could be questioned and that could lead to further investigations and tests. He began to connect highly localised, historically specific, and shifting practices of 'maintenance' in the valley to the more enduring constraints of lake size and water-retention time. This articulation – of things that could change and be changed, together with things that could not be changed or controlled (lake size and retention time) – was not what is normally understood as science, but it did not wholly consist of 'lay expertise' either. Rather, it drew on local *and* scientific knowledges. And this kind of knowing, supported by the mode of engagement of the LCP, enabled many things:

- it alerted the LCP to think about their agency as a collective. Participants began to think about how they could act, and upon which things;
- it sensitised the LCP to the idea that some things (e.g. retention time) were beyond human control;
- it also opened up fertile ground for other connections to be made: e.g., at the 3rd LCP meeting, 32 of 38 participants voted that 'the impact of maintenance issues' should be the 3rd-top priority for the LCP to investigate;[6]
- consequently, at the 5th LCP meeting, a representative of the Environment Agency was invited to speak about the changing regimes of maintenance of the banks of streams ('becks') that flow into Loweswater;[7]
- at the 8th LCP meeting a Loweswater farmer, Ken Bell, informed the LCP of a meeting that had recently been held between five farmers and two members of the National Trust. This meeting aimed to reconsider the management of the becks flowing into and out of the lake. As a result, a number of willow trees which were lying across the outflow of the lake were removed. He described this as constituting a major breakthrough in what had been very poor relations between the two groups (the local National Trust warden and the farmers had previously not been on speaking terms due to mutual blame over the management of the catchment and the subsequent pollution of the lake).[8]

This story highlights the way in which the LCP, through a focus on 'things', and through the creation of a space in which knowledge-based propositions could be made collectively, allowed intimate lay knowledge of the management of things around the lake to cascade and connect to several other forms of knowledge, actors, issues, things, and actions. But the LCP also allowed for contestations to take place in knowledge-making. A second story (given below) highlights the way in which contrasting knowledge claims led to new investigations and a refusal to accept shallow solutions to 'the algae problem'.

---

[6] Minutes of the 3rd LCP meeting: www.lancaster.ac.uk/fass/projects/loweswater/noticeboard.htm.

[7] Talk by Chris Robinson (EA) 'River Maintenance – Changes in Policy' Minutes of the 5th LCP Meeting: www.lancaster.ac.uk/fass/projects/loweswater/noticeboard.htm.

[8] Minutes of the 8th LCP meeting: www.lancaster.ac.uk/fass/projects/loweswater/noticeboard.htm.

Upland Farming in a Highly Connective
Land-Water System

At the 9th LCP meeting in January 2010, the LCP turned its attention to
farming and farm practices in the Loweswater catchment. Even before the
LCP had started in 2007, local residents had expressed their concern that the
new forum would address the issue of upland hill farming: 'I hope that the
problems faced by Cumbrian farmers will be tackled. They have to balance
making a living with any potential dangers to the environment in doing this.'[9]
It was also widely acknowledged by local residents, as one participant had
commented in the very first meeting of the LCP, that farming shapes the very
form of Loweswater: 'When we look out of the window, when we see the fells,
it is what it is because of generations of farmers.'[10] However, despite this
empathy for the livelihoods of farmers in Loweswater, all participants were
acutely aware that farmers had historically been blamed for the algae problem.
Connections had long been made between the intensification of beef-farming
practices, particularly via use of artificial fertilisers on grass crops, and the
nourishment of cyanobacteria in the lake by the element phosphorus (P).
Indeed, ecologists had produced evidence that elevated P levels in the lake
had largely resulted from changes in farming practices (Bennion et al., 2000).

Farmers' routine applications of artificial fertilisers, including P, to the land
were caught up in these sensitive understandings. The LCP, however, had
a mandate to investigate all the 'things' that were connected to lake water
quality. It commissioned a local agricultural consultant, John Rockcliffe, to test
soils on the nine working farms with 'in-bye land'[11] that drained into the
catchment and to carry out a 'nutrient budget' to measure how much P was
needed for the grass crop being grown. Once this level was ascertained, he
would advise the farmers to apply organic and non-organic fertilisers up to that
level, with no permitted exceedance. Rockcliffe carried out his investigation in
the autumn of 2008, coming back to the LCP in January 2009 to report back on
his findings. He reported that the soils of only three farms in the catchment were
judged to have excessive P levels required for plant growth. He also informed
the LCP that the three farmers concerned had seen these results and had
subsequently agreed not to apply any chemical phosphorus fertiliser that

---

[9] Response to a questionnaire sent out March 2007. This response was given in answer to Question
3 'Do you have any comments about the [three year, 2007–2010, RELU] project overall?'

[10] Local resident Alex Bond speaking at the first LCP meeting, June 2008.

[11] In-bye land is grassland that is usually near the farm or homestead and bound by a fence, wall, or
hedge, in contrast to open 'fell'. It is often 'improved' through the creation of drainage and the
application of fertilisers and animal wastes in order to increase the productivity of the grass crop.

same year. It seemed that the problem of P nourishment of cyanobacteria could be easily resolved.

However, when this evidence was aired in the forum of the LCP, another participant, lake ecologist Stephen Maberly, suggested that, in his view, the P problem was not so readily resolved. He explained that, from an ecological perspective, soil and water are highly connected systems, and that, despite Rockcliffe's findings, water quality in Loweswater could still be deteriorating as a result of P inputs into the lake.

As we have seen, the principles or 'rules of engagement' of the LCP suggested that all representations could and should be doubted and questioned. The principles of the LCP also suggested that new connections between people and nature (including phosphorus and cyanobacteria) should be encouraged. Maberly suggested that a new study was needed to further investigate the complex connections between phosphorus inputs on the land and the water quality of Loweswater. Importantly, he suggested that this research might not produce definite answers, but it would give further indications as to what was going on regarding flows of P to the lake. A professional scientist himself, he was advocating the creation of new propositions and testing that could be carried out by his research institute with the help of a local farmer. As in the first story given above, the LCP once again provided a structure through which propositions based on multiple sources of scientific and lay expertise could be made, questioned, and tested. This was a form of knowledge-making that acknowledged different knowledge contributions and built upon them, rather than the traditional mode of science that established the facts of the matter with little room for debate or contestation.

These examples are only a few of the projects that were engendered through LCP discussions, and that supported the hybridisation of scientific and lay expertise.

Conclusion

The principles by which the LCP knowledge collective worked seemed, in some ways, peculiar and counterintuitive. They served to establish a forum where nature is not self-evident, where knowledge and expertise have to be debated, where uncertainties are accepted as the main condition humans are in, where the creation of connections between people and things is highly valued, and where doubt and questioning is extended to all representations. But these principles had the effect of enabling all participants – be they farmers, local residents, scientific agency staff, professional ecologists, etc. – to scrutinise, question, and probe pieces of 'evidence', and to demonstrate their own particular forms of expertise. Such expertise, in turn, would be scrutinised, doubted, and questioned by the collective. But also, crucially, all participants were enabled, through these principles, to build not inscrutable facts produced by science, but propositions and suggestions that could lead to further collective steps towards understanding 'the algae problem' better. By the end of the three years, LCP participants reported that they had become interested in a vast number of interconnected things:[12]

- hydrogeomorphology and future water course management;
- the interaction between people and the environment;
- septic tanks;
- the exploitation of the local farming way of life;
- their community;
- phosphorus in the catchment and its interrelationship with land management practices;
- fish populations;
- land management decisions;
- the locality, its past, its problems;
- analysis of detergents and soaps – advice on what to use;
- ecology, biology, and geomorphology of the area;
- farming and its effect on the lake and the valley;
- history of land use;
- management of becks;
- sediment in the lake and the diatoms;
- and many more . . .

Many of the participants had become highly inquisitive 'scientists', thinking up research ideas, prioritising what the LCP needed to find out, carrying out small projects, critically appraising others' research and results – and thereby blurring discrete notions of scientific and lay expertise in fascinating and lively ways.

This momentum, and its concomitant mode of inquiry and research, has also continued beyond the formal end of the LCP (December 2010). Local residents, including a farmer, subsequently worked together to propose a follow-on project to the UK's Department of Food and Rural Affairs. They gained the requested funding and commenced a further three-year project in

---

[12] The list is taken from the results of a questionnaire given to participants in June 2010 about their views of the LCP. The order in which responses were gathered is the order in which they are represented here.

2012 which they called the Loweswater Care Programme.[13] This project brought scientists, the local West Cumbria Rivers Trust, and local residents together to pool their expertise and steer a programme concentrating on lake-water monitoring and the amelioration of farm infrastructures and technologies for handling and utilising animal farm wastes. The LCP, in effect, was continued, albeit in a modified form and with a tighter focus.

To sum up, what this case study shows is that a somewhat counterintuitive set of principles, underpinning what might look like an ordinary problem-based participatory forum, has been able to support an extraordinary proliferation and diversity of knowledge-making practices that both open out what 'the problem' is and blur the boundaries between scientific and lay expertise in productive ways.

References

Barad, K. (2007). *Meeting the Universe Halfway: Quantum Physics and the Entanglement of Matter and Meaning*. London: Duke University Press.

Bennion, H., Appleby, P., Boyle, J., Carvalho, L., Luckes, S., and Henderson, A. (2000). *Water Quality Investigation of Loweswater, Cumbria: Final Report to the Environment Agency*. London: Environmental Change Research Centre.

Chilvers, J., and Kearnes, M., eds. (2015). *Remaking Participation: Science, Environment and Emergent Publics*. New York: Routledge.

Haraway, D. (1988). Situated Knowledges: The Science Question in Feminism and the Privilege of Partial Perspective. *Feminist Studies*, 14(3), 575–599.

Latour, B. (2004). *Politics of Nature: How to Bring the Sciences into Democracy*. Cambridge: Harvard University Press.

Mol, A. (2002). *The Body Multiple: Ontology in Medical Practice*. London: Duke University Press.

Stengers, I. (2005). Deleuze and Guattari's Last Enigmatic Message. *Angelaki*, 10(2), 151–167.

---

[13] http://westcumbriariverstrust.org/projects/the-loweswater-care-programme.

9

Environmental Experts at the Science–Policy–Society Interface

ESTHER TURNHOUT

9.1 Navigating the Science–Policy–Society Interface

Environmental experts are found in many different places and organisations. They work not only for universities or research institutes, but also for companies, consultancy firms, governmental departments, advisory bodies, and non-governmental organisations (NGOs). As discussed in Chapter 6, they engage in a wide variety of activities, including the undertaking of basic or applied research, as well as offering advice to policy makers or companies, evaluating the effects of specific measures, developing policies, communicating with the media, and many more (Mayer et al., 2004). The different institutional contexts in which experts operate together, the variety of the tasks and activities they undertake, and the responsibilities they have mean that experts have a wide array of options at their disposal in deciding upon their role at the science–policy–society interface. This chapter will present those different options, and will also discuss the specific challenges and opportunities of each of these options. In view of these challenges and opportunities, it is important that experts reflect on the implications of the choices they make.

To be effective as an environmental expert, you will need to think strategically about how you will interact with the potential users of your knowledge and expertise. As earlier chapters in the book have demonstrated, the idea of an expert who delivers objective knowledge that is then translated into policy decisions or solutions is an oversimplification. What kinds of knowledge and expertise will be considered credible, salient, and legitimate depends to a significant extent on the kind of environmental problem that you are dealing with and on the institutional context in which you are situated (see Chapter 6). Moreover, as discussed in Chapter 3, the exact nature of an environmental problem is not pre-defined. It is not self-evident that all parties involved will automatically agree on the nature of the problem, its causes, or the preferred course of action. In view of these complexities, experts need to be able to adapt to the different circumstances in which they find themselves and address the different specific knowledge needs of a diversity of actors.

As outlined in Chapter 4, the different challenges involved when experts attempt to navigate the science–policy–society interface are often interpreted as a dilemma between distance and engagement. When experts remain at a distance from societal and policy actors, the idea is that they are able to remain objective and independent. They can play the role of a pure scientist who conducts research without outside interference. The risk, however, is that their knowledge may not be used. If experts want to ensure the relevance and application of their knowledge, they will have to engage with actors in policy and society. But this is also risky because such engagement might threaten their – perceived – objectivity and independence, thereby undermining their authority and credibility. They risk being seen as issue advocates who are primarily interested in arguing for a specific preferred option or intervention and who use knowledge to justify this preference. In other words, this dilemma between distance and engagement is also a dilemma between authority, which requires distance, and relevance, which requires engagement with knowledge users.

The dilemma between distance and engagement paints a relatively black and white picture in which environmental experts can either retreat into their ivory tower to remain pure scientists, or become – or risk being seen as – issue advocates who use knowledge to advocate for their preferences. Fortunately, the spectrum of possibilities is wider than this, and experts have different options available that will enable them to maintain credibility and authority while at the same time engaging with societal and policy actors and ensuring the relevance of their knowledge and expertise (Turnhout et al., 2013). However, this is not easy and it requires continuous reflection and consideration of the following two questions:

1) How is the problem framed?
2) What and whose values and interests are at stake?

The answers to these questions will inform the different choices that experts can make about what kinds of interactions between producers and users of knowledge are possible and desirable and what role they can assume. We discuss these different options, including their challenges and opportunities, by distinguishing three broad modalities or repertoires (drawing in part on Turnhout et al., 2013 and Turnhout, 2017). The first is the modality of 'servicing'. Here, experts offer their knowledge and expertise in response to questions by commissioners of knowledge such as policy makers or other societal actors. The second modality is that of 'advocating', where experts use knowledge to propose specific options or alternatives. The third modality is called 'diversifying'. In this modality, experts express their commitment to pluralism by addressing the perspectives and knowledge needs of a diversity of actors. By discussing the different challenges and opportunities of these three modalities, this chapter will make clear that each of these can be effective and legitimate in different kinds of environmental problems. Despite the fact that none of these modalities are value free – as Chapters 2 and 3 have demonstrated, all processes of knowledge production involve value-laden choices – they do offer different opportunities for experts to balance credibility, authority, relevance, and engagement. The chapter concludes by foregrounding the ethical dimensions of the science–policy–society interface.

9.2 Servicing

The most traditional role for environmental experts is to do research and supply knowledge that may be used in decision-making processes. This idea corresponds very well with the linear model that was introduced in Chapter 4. In this role, the division of labour between experts and knowledge users is clear and straightforward. Users do not interfere with the knowledge production process, and experts do not interfere with the ways in which decision makers use this knowledge. It is the primary modus operandi for many expert organisations, including, for example, the US Environmental Protection Agency (Jasanoff, 1990) and the Dutch Environmental Assessment Agency (Huitema and Turnhout, 2009). Pielke (2007) has argued that in these organisations, experts often play the role of arbiter who delivers relevant and usable knowledge in response to societal or policy questions.

For this role to be effective, it is important that the question is clearly articulated as a knowledge question so that the arbiter can gather or synthesise knowledge to answer this question in a relatively straightforward way. For example, a question about certain specified environmental impacts – for example, the population or breeding success of a certain bird species – of a specified measure or intervention is one that an arbiter can address effectively. Arbiters stay quite far removed from societal and policy actors. They are interested in supplying usable knowledge, but do not themselves decide what knowledge prospective users should find usable and relevant. And, apart from correcting possible errors or misinterpretations, they will not interfere with the decisions that are made on the basis of the supplied knowledge. With this description, assuming a role as an arbiter appears to be an optimal strategy for experts to maintain their independence and authority while at the same time ensuring that their expertise contributes to environmental issues. However, in practice this is not so simple, for a number of reasons.

The first reason is that in most cases, the initial questions are not always clear. For example, a policy maker is interested in a wide but unspecified range of environmental impacts, or is considering a wide range of possible options. In those cases, more intensive interaction is needed to specify or operationalise the questions before the arbiter can start. Policy options need to be translated from strategic visions or ambitions into concrete and discrete sets of actions, and indicators need to be selected and operationalised in order to make the relevant environmental impacts measurable. However, societal and policy actors often lack the expertise to be able to do this and depend on experts to help them to make these choices. And this is where difficulties come in. To a certain extent, when experts help translate policy ambitions into specific interventions they co-design policies, and when experts help select indicators to specify environmental impacts they help shape the environmental ambition level of policy. Often, doing this involves quite a bit of back-and-forth between the experts, who use their knowledge and expertise to indicate what data, monitoring systems, or models they have available, and the policy makers, who need to check if the proposed translations and indicators still match up with their original questions. As the case about fisheries demonstrates (Case E: Expertise for European Fisheries Policy), available data, monitoring schemes, and models often serve as boundary objects that serve to organise the science–policy–society interface and help structure what knowledge and expertise is relevant, as well as what the scope of decision-making options is.

The second reason is that the role of an arbiter can only be effective when environmental problems are defined and framed as technical problems and when there is consensus about this technical framing. Only in those cases will it be possible to ask questions that arbiters can answer and use the answers to effectively address environmental problems. As explained in Chapter 3, the way in which environmental problems are framed is never self-evident, and there are all kinds of political and normative assumptions behind such a technical framing. Referring again to the fisheries case, the procedure to establish the total allowable catch is based on the idea that fisheries policy should be based on the distribution of single-species quota. Other alternatives are not considered, and they cannot be considered because the only available data are projections from single-species fish models. Similarly, in the example of fracking (Case C: Whose Deficit Anyway?), asking for expertise about the drinking-water contamination risks of fracking implies that these risks are the main factor in assessing whether this technology should be allowed, thereby ignoring not only other potential risks, but also the question of the desirability and necessity of fracking vis-à-vis alternative energy solutions. It can also become quite difficult for these other arguments to be voiced and heard: 'if the technology is safe, why would you be against it?' A lack of understanding on the part of experts and policy makers can subsequently trigger responses in line with the information deficit model discussed in Chapter 4 to better explain the safety of this technology, thereby exacerbating public resistance. In other words, dominant technical framings of environmental problems often exclude other political and social considerations, values, and interests. And this means that when arbiters accept or endorse such technical framing in order to do their work, they are implicitly – and perhaps often also unwittingly – endorsing the political and social considerations, values, and interests that underwrite this technical framing.

These two reasons make clear that although the role of arbiter seems – and often is – a safe and efficient way for experts to balance distance and engagement, it is not value neutral or interest free. Accepting a role as an arbiter means that you accept the norms, interests, and values that informed the question you are asked to answer. Although arbiters may not actively advocate for these in the way that issue advocates do, this does not mean that their work is free of them. While this agreement between arbiters and policy makers about what the problem is and what knowledge is relevant for the arbiter role makes for smooth and effective interactions between knowledge production and use, it also entails a strengthening of certain norms, values, and interests and the exclusion of others. It is important to stress that this can be legitimate in specific cases, when the technical nature of the problem is not in dispute or when the norms, interests, and values served are considered legitimate. However, it is important for arbiters to reflect on whose interests are served by answering the questions of what is being excluded, and how this can be justified.

9.3 Advocating

It can also be legitimate for environmental experts to produce knowledge that fits with their own norms, values, and interests – rather than accepting those of the commissioners of knowledge like arbiters do. For example, experts in ecology or conservation biology are often motivated by a concern for nature and are committed to contributing to conservation. This influences the topics that they do research on, and it is not uncommon for these experts to use their findings to propose recommendations for policy and management. Within the social sciences – for example, in development studies or anthropology – experts often play a similar role in investigating and advocating for the needs of marginalised actors.

When experts connect research findings to particular preferred courses of action, they assume the role of an issue advocate. This role is typically found in NGOs, but also in so-called curiosity-driven or basic research, where experts are free to identify their own research topics, questions, and methods.

In many cases, the role of the issue advocate is not very problematic. It is acceptable that experts from NGOs use knowledge to argue for their case, and in curiosity-driven research it is accepted that scientists choose their own topics and research questions to fit with their own problem framings and preferred solutions. Particularly when the technical framing of problems is not disputed, experts are even expected to use their knowledge to formulate recommendations to improve a certain state of affairs. As such, in most cases, issue advocacy is not so far removed from accepted practice in science. However, this changes in the case of knowledge controversies. When there is no agreement about the framing and definition of the problem, about what knowledge and expertise are relevant, or about how this knowledge should be interpreted and used in decision making, issue advocacy becomes much more visible and risky. In those cases, opponents in a controversy can easily discredit experts for being interested and politically motivated when they use knowledge and expertise to articulate recommendations or suggest preferred policy options. As discussed in more detail in Chapter 4, in environmental controversies experts and their knowledge can easily become part of the debate.

Within science, issue advocates have a bad name because they are often associated with the manipulation of scientific knowledge for political purposes. Certainly, there are cases where this happens (e.g. Oreskes and Conway, 2010), but this does not mean that issue advocates necessarily *misuse* knowledge and expertise to argue for certain preferences. It should be recognised that different problem framings inevitably allow for different legitimate interpretations of knowledge – of what knowledge is relevant and of how knowledge should inform decisions. Moreover, the role of issue advocate is indispensable in democratic societies (Pielke, 2007). There are instances where it is crucially important for experts to assume the role of issue advocate in order to open up and criticise certain dominant problem frames or to strengthen marginalised actors and perspectives. Although this strategy can be risky, since accusations of advocacy are often damaging for a person or a scientific career, it is clear that the role of issue advocate is not necessarily more political or less neutral or value free than the role of arbiter discussed in the previous paragraph.

9.4 Diversifying

In situations where there is no consensus about the problem framing, it is difficult for experts to avoid being seen as an issue advocate because the values that are inevitably associated with knowledge will be foregrounded and used to discredit the experts. In those cases, experts can consider a different strategy and assume the role of a broker (Pielke, 2007). This role involves an explicit commitment to diversity and to meeting the needs of different relevant stakeholders. Brokers will use a broad spectrum of perspectives and values to produce knowledge and expertise and to develop and/or evaluate different options and alternatives.

Unlike arbiters, brokers will not uncritically accept a specific question when this question is derived from a narrow and technical problem framing. Instead, they will open up the question to include the knowledge needs of a diversity of relevant stakeholders that may have different ways of framing the problem. Instead of answering the question 'what are the effects of the expansion of this dairy farm on the population size and breeding success of the black tailed godwit [a species of meadow bird]?', a broker will develop different land-use options – including, for example, organic farming, conservation, or recreation – and evaluate these for a diversity of environmental, social, and economic effects. Unlike issue advocates, brokers will not advocate for one specific land-use option and use knowledge about godwits to justify this preference.

The role of broker takes considerable effort to realise. The commissioner of knowledge needs to be open to the idea of generating a diversity of knowledge and options, and the broker needs to invest time and resources to explore what perspectives are relevant to include and subsequently to generate knowledge and identify possible options based on these perspectives. Although the broker tries to avoid taking sides in a debate through a commitment to diversity, this role is not value free – diversity is itself a value – and it involves risks for the authority and credibility of experts. As the case about CO_2 capture and storage illustrates (Case J: Groupthink and Whistle Blowers in CO_2 Capture and Storage), simply arguing for the need to include multiple perspectives, or attempting to open up certain dominant problem framings, can result in criticisms and accusations of being an advocate for these perspectives.

When brokers, instead of only exploring the perspectives of stakeholders, also involve them in the research process, their role changes into that of a participatory expert (Turnhout et al., 2013). Using the previous example, a participatory expert will organise and facilitate a participatory process in which experts collaborate with stakeholders to identify possible alternative land-use options and evaluate those in terms of the different criteria that are considered relevant. This role can also be challenging not only because of the time investment, but also because of the broad spectrum of skills required. This is not only about producing knowledge and communicating it in an understandable way; participatory experts need to be able to organise and facilitate processes of engagement and knowledge production, and they need to be able to deal with a wide variety of perspectives, ideas, interests, and values. A second reason why this role can be challenging for experts is because the products of these activities are not just the result of scientific knowledge and methodologies, but are co-created by stakeholders, who bring in their own local knowledge and expertise. Thirdly, since participatory experts start from a societal issue or problem to co-create knowledge-based solutions, they bring knowledge and action closely together to the extent that they become difficult to disentangle. As such, the products of these processes are not always clearly recognisable as knowledge. This is also more generally true for the participatory knowledge production processes (see Chapter 8).

Thus, although both brokers and participatory experts are committed to including a diversity of perspectives and values, there are two important distinctions. First, participatory experts consider stakeholders not just as having relevant values and perspectives – as the broker does – but also as holders of relevant and credible knowledge. As such, they blur the distinction between scientific and other forms of knowledge and expertise (see Chapter 8). Second, knowledge and action are seen as intertwined in the context of a specific societal issue or problem. As such, they fundamentally break away from the linear model, and the separation between knowledge production and use that this model implies.

9.5 Expert Roles and Dilemmas

The discussion of expert roles makes clear that at the interface between science, policy, and society, experts have different possibilities and options that go beyond the stalemate dichotomy of being either a pure scientist or a politically motivated issue advocate. It has also made clear that none of these roles – including that of the pure scientist – are value free. Nevertheless, the different roles all – perhaps with the exception of the pure scientist – offer opportunities to maintain credibility and authority while at the same time engaging with societal and policy actors to make sure that knowledge is usable and relevant. Each role also comes with particular challenges that require ethical reflection. Such reflection can be structured using the two questions posed in the first section:

1) How is the problem framed?
2) What and whose values and interests are at stake?

When a problem is framed in a narrow technical way, experts should reflect on whether this dominant framing is legitimate. Is there consensus about this problem framing and can it be justified, or does it exclude relevant values, interests, and perspectives? In cases where this framing is considered legitimate, experts can assume the role of arbiter. They must, however, be aware of the problem framing that they are working in and supporting, and they must recognise that in doing so they are supporting and strengthening specific dominant elite values and interests. When this dominant problem framing is not deemed legitimate, the modality of servicing is problematic. Experts can attempt to challenge this dominant problem framing by advocating for the importance of considering alternative problem framings and of recognising the values, interests, and perspectives of marginalised actors. When there is no consensus about the problem framing, experts can employ the modality of diversifying by bringing the different perspectives in the controversy into productive dialogue and developing options and alternatives for action on the basis of these perspectives. These latter two modalities are more risky than that of servicing because they require engagement in controversies or with actors outside the elites. Consequently, it is more likely that values and perspectives will not remain implicit or invisible and will be used to discredit experts and expertise.

By now, it will be clear that choosing among roles and modalities is not necessarily about choosing to engage or not, or to include values or not, since the engagement of values cannot be avoided. Instead, the important question becomes what and whose values are promoted in knowledge and expertise, and how that can be justified. Explicitly addressing this difficult question is important to assess what roles are legitimate in what situations. However, it is also important to recognise that the processes discussed here are dynamic. Powerful actors may succeed in establishing a technical framing of an environmental problem and ignore or neutralise other problem definitions. However, this situation may not last, and marginalised actors may be able to successfully challenge this framing after which a controversy emerges. The expert who thought he could unproblematically work as an arbiter is now faced with a new situation and must reconsider his role. Given these dynamics, it is important to be flexible. The roles outlined in this chapter are not intended as fixed boxes, but aim to sketch a spectrum or repertoire of possibilities. Roles do not refer to individuals in the sense that they define the intrinsic identity of an expert. Nor are they automatically linked to certain types of institutions in the sense that universities harbour pure scientists only while consultancies harbour arbiters. Experts may choose to alternate between roles, or may even perform their own mixtures. Dealing with environmental issues requires experts to interpret the situation they are in, to adapt to changing circumstances, to reflect on risks and opportunities, and to reconsider their activities and strategies accordingly.

In practice, the distinctions between different expert roles are not always easy to recognise. The term 'knowledge broker' is used in the literature and in practice to denote a wide variety of roles and activities at the science–policy–society interface, potentially encompassing all three modalities (servicing, advocating, and diversifying) discussed in this chapter (Michaels, 2009). More generally, the point of this chapter is not to offer an exhaustive classification of clearly demarcated roles, but rather to outline a diversity of options. Yet, recent studies suggest that the creation of connections and the co-production of actionable knowledge – that is, knowledge that is situated in practice and catalyses concerted action – are greatly facilitated by explicit attention to safeguarding diversity of knowledge as well as of options (Rantala et al., 2017; Bednarek et al., 2016; Brunet et al., 2018, see also Case I: The Loweswater Care Project). This resonates well with Stirling's (2010: 1029) argument that despite a common belief that decision makers require unequivocal and straightforward, often quantitative, knowledge, the role of knowledge can become more robust and effective when it explicitly recognises the 'intrinsically plural [and] conditional nature of knowledge'.

9.6 Ethics and Integrity at the Science–Policy–Society Interface

The different dilemmas and ethical considerations discussed in the previous section are only partly up to individual experts to reflect and act upon. In practice, expert roles are shaped in the process; in the interactions between the expert and the prospected users of knowledge and expertise. This means that these users also determine the roles that can and cannot be performed. In addition, the organisations that experts work for also have specific expectations about what their experts should and should not do. Partly this is about practical matters such as time and costs. In particular, the processes required for the role of brokers or participatory experts are often time consuming and costly, and this may not fit with the expectations of the commissioner or of the own organization. Incentive structures may also work against these roles when experts are primarily evaluated on the basis of scientific output and are not rewarded for working with stakeholders. But there are also other factors that determine what roles will be possible. Dominant stakeholders may not understand or agree with the need to include different perspectives and stakeholders, and they may not appreciate it when experts argue for this. Second, established scientific norms of objectivity or neutrality (see Chapter 4) may prevent scientists from actively engaging with stakeholder perspectives and values, or from including lay expertise on an equal footing with scientific expertise. Even if, as the chapters in this book have demonstrated, these norms are misguided, they are strongly institutionalised in many expert organisations and universities and, consequently, they offer strong incentives for experts to stick with the relatively 'safe', yet not value-free, roles of the pure scientist or the arbiter.

This means that ethics is not just a matter of individual reflection and behaviour by experts, but also a matter of institutions and relationships (also see Turnhout, 2017). Current codes of conduct for scientific ethics and integrity, although important, are not sufficient because they mostly address individual behaviour in the execution of research. However, as this chapter has demonstrated, integrity and ethics should not only be about 'doing things right' but also about 'doing the right things' (Drucker, 1993). Expert integrity and ethics, then, are about creating the conditions to make sure that this second question – are we doing the right things? – can be asked, and that experts can act upon the answer to this question. Sometimes this will involve not accepting a certain commission, or accepting a commission that certain other powerful relations of the organisations do not appreciate. Sometimes it will involve negotiating what questions to address, what stakeholders to involve, or what methods for (participatory) knowledge production to apply. Not all questions deserve expert input, and sometimes experts need to challenge dominant frames.

References

Bednarek, A. T., Shouse, B., Hudson, C. G., and Goldburg, R. (2016). Science-policy Intermediaries from a Practitioner's Perspective: The Lenfest Ocean Program Experience. *Science and Public Policy*, 43, 291–300.

Brunet, L., Tuomisaari, J., Lavorel, S., et al. (2018). Actionable Knowledge for Land Use Planning: Making Ecosystem Services Operational. *Land Use Policy*, 72, 27–34.

Drucker, P. (1993). *Management: Tasks, Responsibilities, Practices*. New York: Harper Business.

Huitema, D., and Turnhout, E. (2009). Working at the Science–Policy Interface: A Discursive Analysis of Boundary Work at the Netherlands Environmental Assessment Agency. *Environmental Politics*, 18(4), 576–594.

Jasanoff, S. (1990). *The Fifth Branch: Science Advisers as Policymakers*. Cambridge: Harvard University Press.

Mayer, I., Daalen, C., and Bots, P. (2004). Perspectives on Policy Analyses: A Framework for Understanding and Design. *International Journal of Technology, Policy and Management*, 4(2), 169–191.

Michaels, S. (2009). Matching Knowledge Brokering Strategies to Environmental Policy Problems and Settings. *Environmental Science & Policy*, 12, 994–1011.

Oreskes, N., and Conway, E. (2010). *Merchants of Doubt*. London: Bloomsbury Press.

Pielke, R. (2007). *The Honest Broker*. Cambridge: Cambridge University Press.

Rantala, L., Sarkki, S., Karjalainen, T. P., and Rossi, P. M. (2017). How to Earn the Status of Honest Broker? Scientists' Roles Facilitating the Political Water Supply Decision-making Process. *Society & Natural Resources*, 30, 1288–1298.

Stirling, A. (2010). Keep it Complex. *Nature* 468, 1029–1031.

Turnhout, E. (2017). Integere relaties: wetenschappelijke integriteit en de verhouding tussen wetenschap en samenleving. *Beleid en Maatschappij*, 44(1), 58–66.

Turnhout, E., Stuiver, M., Klostermann, J., Harms, B., and Leeuwis, C. (2013). New Roles of Science in Society: Different Repertoires of Knowledge Brokering. *Science and Public Policy*, 40(3), 354–365

Case J Groupthink and Whistle Blowers in CO_2 Capture and Storage

HELEEN DE CONINCK

This case of CO_2 capture and storage (CCS) is a detailed example of an expert who stepped out of an arbiter or honest broker role and took on a more engaged and critical stance. The arbiter no longer found the terms of advice acceptable and felt the need to question the assumptions in the assessment of carbon capture and storage as a means to mitigate climate change. In her eyes, the CCS experts were turning into issue advocates, pushing their preferred solution. In the Netherlands, the issue came to a head in the media over a plan to store CO_2 deep underground in Barendrecht, a town near the Dutch city of Rotterdam.

Introduction

This case illustrates the role of experts in a controversial and salient debate. The case in question is my involvement in the debate on the climate change mitigation technology of carbon dioxide capture and storage (CCS) in the Netherlands in the years 2009 and 2010. In this case, I reflect on my own role as an expert (or perhaps a whistle blower) around a technology. This account will unavoidably be coloured by my own perceptions of what was for me an exhilarating, stressful, and educational time.

I will start with a short description of global CCS developments over time and my roles in those times. I will then turn to the CCS community in the Netherlands and what happened around the planned CCS project in Barendrecht, which ended up being cancelled in 2010 due to strong local public resistance, and my (small) role in those discussions. I will reflect on how I fulfilled my expert role (as explained in Chapter 9). Finally I will try to explain my own actions and those of other CCS experts by referring to the poor structuring of the problem and the limited engagement between those inside the CCS coalition of industry, government, and research, and those outside.

A Short History of Carbon Dioxide Capture and Storage

Most climate change mitigation options reduce emissions of greenhouse gases into the atmosphere by avoiding the production of those greenhouse gases. CCS is different as it reduces the emission of carbon dioxide to the atmosphere by capturing the gas from the exhaust gases of a CO_2 source and storing it permanently deep underground. Suitable CO_2 sources need to be stationary and large scale, and the higher the CO_2 concentration, the easier the capture of CO_2. Examples of suitable sources of CO_2 include refineries, coal- or gas-fired power plants, steel plants, cement factories, and chemical plants.

The underground storage is intended to be in geological formations, such as depleted oil or gas reservoirs or saline aquifers (salty brines), more than 800 metres below the surface. The CO_2 is injected in a supercritical phase, having the volume of a fluid but other characteristics of a gas, and, at depths greater than 800 metres, the pressures and temperatures are such that CO_2 will remain supercritical. Before the CO_2 is injected in the underground, it is compressed and transported, also under pressure, in a pipeline or by ship, depending on the distance, quantity, and whether the transport will be over land or sea.

Capturing CO_2 involves building a small chemical plant, which uses a significant amount of energy; this is the most costly step in the CO_2 capture, transport, and storage chain. The transportation of CO_2 happens in a large man-made structure and is estimated to bear the greatest risk of accidents and leakage, but it is also the element in the chain that is currently best understood as there are already thousands of kilometres of CO_2 pipelines, in particular in North America. Storage of CO_2 is a fairly well-known practice, thanks to decades of experience in injecting fluids underground in the oil and gas industry. Nevertheless, this step causes most concern with many people. Experts consider the risk of leakage of CO_2 from geological reservoirs to be very low if the reservoirs are adequately selected, monitored, and managed (IPCC, 2005). However, this 'expert' claim is contested by sceptics of CCS technology, and was one of the controversial issues around the Barendrecht CCS project.

CCS is considered to be an attractive option because it allows for continued use of fossil fuels in power production, industry, and perhaps even transportation while significantly reducing the associated CO_2 emissions and, consequentially, climate impacts. It is also said that it would avoid the need to make difficult changes in behaviour, a popular argument among engineers who have more confidence in technological solutions than in influencing human behaviour.

CCS emerged on the policy agenda in the first half of the 2000s. This was a time of low energy prices, when development of energy efficiency and renewable energy technologies was slow and it seemed increasingly unlikely that these options alone would deliver the CO_2 emission reductions needed to avoid dangerous climate change. As a result, CCS gained increasing traction as an attractive solution. Various countries, as well as several oil and gas companies, had been investing in research and development around CCS technology since the early 1990s, with the first demonstration projects getting off the ground, in particular in Norway, in 1996 and the early 2000s. One of the International Energy Agency implementing agreements, its Greenhouse Gas R&D Programme, played an important role by synthesising research results, sponsoring a biannual conference that attracted increasing attendance, and advocating CCS.

In 2001, the United Nations Framework Convention on Climate Change (UNFCCC) requested that the Intergovernmental Panel on Climate Change (IPCC) prepare a technical paper on CCS. After a scoping meeting in 2002 in Regina, Canada, the IPCC decided to prepare a Special Report on Carbon Dioxide Capture and Storage (SRCCS). This 250-page report, which was finalised in 2005, assessed and summarised research on capture, transport and storage technology, risks, environmental impacts, costs, and policy and public perception aspects of CCS. It contained a full chapter on CO_2 storage in the oceans (as opposed to in the geological underground), which essentially led to the abandonment of that storage option because of limited storage effectiveness and significant environmental impacts. It also concluded that the potential for reuse of CO_2 in different applications would have only a limited impact on CO_2 emissions. The main conclusion and outcome of the SRCCS was that carbon dioxide capture and geological storage could make a significant contribution to climate-change mitigation, and that costs would be generally lower than many other mitigation options.

The SRCCS catalysed a flurry of activities by various governments to enable CCS. The European Union, which kicked off its Emissions Trading Scheme (ETS) in 2005, started efforts to make CCS compatible with its ETS, which at that time featured carbon prices that seemed sufficient for CCS projects. Various other countries, such as Australia, Canada, and the United States, started policies to help move CCS forward. Parties in the UNFCCC were pleading for inclusion of CCS in the Kyoto Protocol's Clean Development Mechanism (CDM), so that developing countries could also get carbon credits for CCS projects. Various international organisations got involved, such as the Organisation for Petroleum Exporting Countries (OPEC) and the World Bank (De Coninck and Bäckstrand, 2011).

My Work on CCS

I was an engaged participant in the global CCS discussions during most of the above-mentioned period. The first time I heard about CCS was in 2001, when I applied for a PhD position on CCS. I didn't get the job, but a few months later I started working as a junior researcher in international climate policy at the Unit Policy Studies of the Energy Research Centre of the Netherlands (ECN), in the Global Change group. My work mostly involved studies and capacity building related to the then-emerging CDM and other aspects of the Kyoto Protocol and the UN climate negotiations.

One of the additional assignments that came in was to assist the IPCC Technical Support Unit of Working Group III on Mitigation, which at that time was hosted by the Netherlands Environmental Assessment Agency (then part of the RIVM, later called the MNP, and currently known as PBL) as it was co-chaired by Bert Metz. Bert Metz and Ogunlade Davidson of Sierra Leone had just been re-elected as co-chairs of Working Group III and were responsible for the Fourth Assessment Report of the IPCC. As the work on the scoping meeting on CCS had to start promptly, and there was limited capacity in RIVM internally to do this work, I was engaged as a junior support person, joined by a senior staff member of ECN (Daniel Jansen, a chemical engineer specialised in carbon dioxide capture technologies).

From 2002 to 2005, I devoted about half of my working time to supporting the SRCCS. For the other half, I remained involved in work around international climate policy, in particular around the CDM and post-Kyoto climate policy. The work on the SRCCS included inviting authors to contribute to reports and answering their questions, the organisation of four Lead Author meetings, the design and management of the review process (which on this first occasion only was done anonymously), editorial work, writing and editing the summaries, and secretarial and other kinds of assistance work during the official approval process of the Summary for Policymakers (SPM). Such line-by-line approval processes of the Summaries for Policymakers of IPCC reports are full-blown UN negotiation processes, with the majority of countries represented in a plenary meeting that lasted for four days and where negotiations on contentious topics went on well into the night.

I also represented the IPCC on a number of occasions, during conferences and meetings of international organisations where information on the report was to be given or at which the IPCC wanted to be represented. For example, just two days after the SRCCS SPM approval in September 2005 in Montreal,

I presented the report at a member's meeting of the Carbon Sequestration Leadership Forum in Berlin. It should be said that my role was purely supportive: the substance of the report was provided by the author team, and the final responsibility of the report and its summaries rested with the Working Group III co-chairs, and eventually with the governments represented on the IPCC Plenary.

During my time with the IPCC, I was deeply involved in the CCS community. Though not a researcher on CCS at that time, as a member of the technical secretariat I needed to have a deep understanding of all aspects of CCS. The author team of the SRCCS were the real experts and scientists, but I became knowledgeable on everything from energy modelling to CO_2 capture and from policy processes to underground storage thanks to the excellent access I had to the 100+ authors from all over the world. I asked them a lot of questions, and they were happy to help me as I helped the co-chairs understand, which in turn allowed them to do a better job. However, at the same time, in order for the report to be good, I needed to take a critical stance. If the SRCCS were to oversell CCS, it would harm the credibility of the IPCC, which is supposed to refrain from policy advice and technological bias.

It was not until after the completion of the SRCCS that I started to do research on CCS myself. In my research, I connected my experiences in the SRCCS process with new research questions in the field of climate policy. I worked on questions around CCS in the CDM and on EU policy on CCS. With ECN, I participated in an EU FP6 project called ACCSEPT, which looked into regulatory, legal, public perception, and economic aspects of CCS (De Coninck et al., 2009). I helped coordinate the project that drafted the Directive on the Geological Storage of CO_2 for the DG Environment of the European Commission (European Commission, 2009). I worked on various research and capacity-building projects, generally in the role of a connector between science and policy. However, in parallel I remained a researcher at ECN, and I continued to consider myself as primarily a scientist. I was at that time also working on a PhD thesis in the field of technology and international climate policy in combination with my work at ECN. I had my own views on CCS, which essentially were that I considered it a necessary technology if climate change was to be prevented. Hence, I could be considered a CCS supporter. However, I disliked it if people called me a CCS advocate; I regarded the technology critically, especially a few years after the SRCCS was published.

An Advocacy Coalition Forms While CCS Is Under Pressure

In the meantime, the CCS work at ECN, my institute, flourished: besides the policy analysis work that several colleagues and myself were doing, a large group of engineers was working on improving CO_2 capture technology for industry and research funders, and a field of work around public perception and stakeholder engagement emerged. Within the Netherlands, the community grew even further. The R&D project CATO (for CO_2 *Afvang, Transport and Opslag* – capture, transport, and storage in Dutch) played a central role in unifying industry, government, and the research community behind a favourable narrative of CCS. At its peak in 2008, the Dutch CCS conference attracted 440 participants, from industry, research, governments, and NGOs.

It was at this event that my discomfort with the CCS community became difficult to ignore. In the IPCC, I had already recognised that some authors had become so involved in CCS that they became advocates and had to be managed very carefully to assess the literature in as neutral a way as possible. Managing the risks advocates posed to the reputation of the IPCC was probably the hardest – and also most intellectually challenging – part of my work at the IPCC. In hindsight, I believe the IPCC did a reasonable job, but we were probably all (including myself) slightly carried away by the promise of low costs and the seemingly good-looking technological maturity of CCS. When the discussions on CCS in the CDM started, the relentless advocacy of several industry stakeholders in the UNFCCC to have developing countries accept CCS, a technology of which they had only recently first heard, also made me realise that there were some large interests behind CCS.

During the 2008 Dutch CCS conference, hardly a critical word was uttered on CCS. When a cheque for €5 million was offered as a pre-payment to the CATO-programme to pursue its activities in a second phase with more than double the funding of the first phase, there was cheering all around. The single Greenpeace representative – probably the only person in the room opposing CCS – must have felt very uncomfortable among so much cheerful and hopeful agreement. I remember that he left during the lunch break.

However, during those years, CCS was already under some pressure. Compared to when the SRCCS was published, several cost factors for CCS (in particular energy and steel prices) had risen dramatically. Consequently, the cost estimates in the SRCCS, and accordingly the projected role of CCS in global energy scenarios, quickly became outdated. When, in the summer of 2008, the financial crisis hit, capital became hard to come by and CO_2 prices went down, further harming the CCS business case. When finally, in December 2009, the Copenhagen climate conference did not deliver much hope for strong international climate action, prospects for CCS looked dim and the mood started turning sour.

Currently (in 2018), CCS is kept alive by several projects in North America that are profitable through Enhanced Oil Recovery (a technique that uses the CO_2 to mobilise more oil out of reservoirs with declining production), smaller demonstrations, and funding by a few governments, in particular Norway and Australia. The large-scale demonstrations planned by the European Union have all been cancelled, the last one (the Rotterdam Opslag and Afvang Demonstratie in the Netherlands) only after a long delay of the final investment decision, because low CO_2 prices have a detrimental effect on commercial viability, even with €330 million in subsidies from the EU and the Dutch government combined. CCS is now included in the CDM, but only got approved in 2011, when the CDM market was already close to collapsing.

The Spotlight on the Barendrecht Project

After the SRCCS demonstrated the relevance of CCS for mitigation, one of the activities of the Dutch government to further CCS in the Netherlands was the funding of demonstration sites in the so-called CRUST programme. The initial proposals included a 2007 submission by Shell for a relatively small subsidy of €30 million. The project proposed compression of 0.7 $MtCO_2$ annually of the already purified CO_2 from a hydrogen facility at one of Shell's refineries in the Rotterdam harbour area, transportation of the CO_2 along an existing gas pipeline, and storage of the CO_2 in two recently depleted gas fields under the suburban area of Barendrecht. The costs of this CCS project were very low because of the high CO_2 concentration, the short transport distance and the presence of existing pipeline corridors, and the state of the depleted reservoirs, which were already well documented and had the well infrastructure still in place. For all involved – in particular the Dutch government and Shell – it seemed that the project only had benefits.

Feenstra et al. (2010) and Van Egmond and Hekkert (2015) provide full accounts of what happened around the Barendrecht project. I will briefly summarise these developments here. After a few information sessions that were mainly conducted by Shell, inhabitants of Barendrecht raised concerns about the safety of the project. The credibility of those involved in the project, in particular Shell but also the Dutch government, was called into doubt. The results of the Environmental Impact Assessment (EIA) were questioned on the basis that the EIA was funded by Shell (although this is normal procedure in such assessments). Local inhabitants had a number of different reasons to oppose the project. They had doubts about safety. They felt that they were being treated as guinea pigs for a dangerous, experimental, and environmentally harmful technology, and they indicated that they already had more than their share of large infrastructural projects in their surroundings, including a major freight rail link and a new highway exit. They felt insufficiently compensated and did not get answers when they asked what would happen if house prices dropped as a consequence of the CCS project. As a result of this opposition, the city council, which was initially in favour of the project, changed its mind. A subsequent strategy by the Dutch government to elevate the decision making to the national level through an economic recovery law to counter the economic crisis backfired: it was seen as further procedural injustice imposed on the people of Barendrecht. When the two ministers responsible – Jacqueline Cramer of the Ministry for the Environment and Maria van der Hoeven of the Ministry of Economic Affairs – organised a meeting in early 2009, in the local theatre, to explain their policy, they met with vehement and highly emotional opposition.

I was not formally involved in the Barendrecht project. Through talking to my colleagues at ECN who were working on public perception of CCS, I knew that those colleagues were eager to conduct surveys and assist with public engagement around the project, and had made attempts to interest the Dutch government and Shell in their assistance and insights early on. However, the project proponents had indicated that they were not interested in their services. At one point (in 2009, I believe), I received a phone call from the Ministry of Environment, asking whether I would be able to conduct a second-opinion EIA, as I was considered relatively neutral on CCS. Although I have a Master's degree in environmental sciences, I felt I had to refuse as I had never led an EIA before and felt I didn't have sufficient knowledge of geological aspects of CO_2 storage to credibly assess its safety. An international group was then formed to comment on the EIA, in which I was not involved.

This phone call from the Ministry, in addition to the earlier-mentioned Dutch CCS conference, made me realise that a problem had arisen in the CCS community. The community had developed into a tightly knit group of academic and applied researchers, policy makers, and people working in various business sectors: service companies, manufacturers, oil and gas companies, and electricity companies. This closeness had as an advantage that it led to a strong, coherent message to policymakers. However, groupthink is a known pitfall of such organisations: as a community, the group may become blind to drawbacks, listen selectively, and dismiss certain types of criticism. In particular, if such criticism, for example about the safety of CCS, is voiced by lay people or those outside of the CCS community, it can be easily dismissed: they don't know what they are talking about, we just need to inform them better and convince them. In my view, this attitude towards local inhabitants contributed to the situation in Barendrecht: Shell and the Dutch government embarked on a strategy to inform and convince, but did not genuinely listen to the concerns voiced. They were not ready to be convinced by the protesters. As a consequence, the people of Barendrecht did not feel that their concerns were taken seriously, and hence they intensified their protests.

In 2009, I was called by a journalist employed at *Nature*, who wanted some general information on CCS and with whom I shared my concerns. He thought I raised some interesting points, and although he did not include it in the piece he was writing, I was later invited to write a short piece in *Nature* on groupthink in the CCS community (De Coninck, 2010). This then was taken up by Dutch newspaper *De Volkskrant*, leading to an interview in the Science section in 2010,[1] and a brief appearance in an item of the Dutch television show *Zembla* on the Barendrecht project. The *Volkskrant* interview and the *Zembla* item appeared within a few weeks of each other in March 2010. Most of the criticism on the Barendrecht project in the Zembla item was uttered by others, who bore the brunt of the attention and criticism from the CCS community. My remarks were only about the community and my observation of groupthink, and not about CCS or about the Barendrecht project, as I did not feel qualified to talk about the latter.

The interview with *De Volkskrant* presented me as a sceptic of CCS and a whistle blower in the CCS community. In an earlier draft of the interview, I had indicated that I wanted a number of modifications to the text, but the journalist had not accepted all of them, especially the ones that were, frankly, a bit nasty about my CCS colleagues. The journalist claimed he thought that my employer, ECN, wanted those changes and that they were not my own view. However, he did not check that with me. In fact, both the management at ECN and my ECN colleagues working on CCS have always been very supportive and encouraged me to make those statements, so the suggestion that I had been censored by ECN was not true.

After the interview, trouble started. Without my knowledge, the *De Volkskrant* journalist had contacted a professor at Utrecht University, a long-established expert on CCS, and discussed my statements (before my comments) with him prior to publication. The professor had been one of the organisers of the first international CCS conference in 1992, and is considered

---

[1] www.volkskrant.nl/nieuws-achtergrond/onenigheid-over-co2-opslag~ba9bafc9/

an authority in the field. Of course, he was disgruntled about my rather blunt allegations (this was based on the version that I had not approved) that the CCS community suffered from groupthink, and said as much to the journalist. The result was that, when my interview appeared, the front page of *De Volkskrant* contained an additional item entitled 'Disagreement on CO_2 storage' with quotes from me and the professor disagreeing[2].

The newspaper publications coincided with a meeting of a committee that organised the next international CCS conference, which was going to be in Amsterdam. The professor translated the *De Volkskrant* interview for the other committee members present, many of whom I had worked with during my IPCC years. Apparently, there was an explosion of indignation in the meeting. He also sent an email to me, copying the ECN directors and several colleagues, in which he argued why I was wrong and why my statements in *De Volkskrant* were out of line, and in which he suggested that the conference committee members all agreed with him. He also questioned my CCS expertise. Again, my ECN colleagues were very supportive, and his copying-in of my superiors had no negative consequences for me. I replied on content and left it at that, although I did have a sleepless night over this. In addition to a few critical remarks, I mostly received support for what I said in *De Volkskrant*.

It was interesting that the professor's response was a textbook example of what often happens in such coalitions: the problem of groupthink, as well as the possible shortcomings of CCS, are denied, the whistle blower is discredited, and his or her expertise is questioned. Those in the coalition respond to critics by exclusion. In this case, the situation was exacerbated by the journalist misquoting me.

Expert Roles in CCS

In the CCS debate, I turned out to be something of a whistle blower who tried to ask critical questions about CCS and the CCS community. While I worked at IPCC, I mainly tried to play a role as arbiter or broker, as that is essentially the IPCC's mandate. Many experts in CCS moved, during and after the IPCC process, into the role of participative experts who work with stakeholders to design and implement solutions. I have also taken that role for a short while, for instance by providing capacity-building services in developing countries, by helping to get CCS legislation in the EU accepted, and by assisting with the inclusion of CCS in the CDM. However, I have always positioned myself in an honest broker, a pure scientist, or an arbiter role.

---

[2] www.volkskrant.nl/wetenschap/onenigheid-over-co2-opslag~b164e168/

My critical attitude was the result of a growing unease about the community in the Netherlands (as described above), but also of the large fossil-resource interests and huge money flows that circulated around CCS. For instance, in 2010 I was invited to the Doha Carbon and Energy Forum, a meeting staged by the government of Qatar to consider an energy future for that small yet super-rich oil and gas state. I went, to find out whether Qatar was really serious about making its energy system more sustainable. The organisers flew in some 200 of the world's leading experts in energy technologies and policy (business class of course), put them up in the Doha Ritz Carlton hotel, and allowed them to talk about climate and energy, moderated by a UK-based meeting facilitation company. What I had feared turned out to be true: the meeting was mainly held to make Qatar's non-extant climate policy look like something, and to legitimise the perpetuation of fossil interests. CCS was used as an excuse to not act on other mitigation options and to continue business as usual. The excessive budget for the meeting and the meagre, fig-leaf-like results were yet another eye-opener. I also became more critical thanks to a group of social scientists working on a special issue of *Global Environmental Change* (Bäckstrand et al., 2011) who involved me yet did not ask the question 'how do we get CCS implemented responsibly?'; instead, they asked 'is CCS a good idea in the first place?'

All in all, I conclude that some aspects of CCS started to conflict with my personal values around how addressing climate change ought to lead to a more equal and sustainable society. Many of the CCS experts I met acted from a deep conviction that climate change needs to be resolved, and that CCS is indis-pensable to accomplish this. Others were more affiliated with the money-washed world of the oil industry. Once I became convinced that CCS had formed an industrial complex, a de facto coalition of industry, government, and scientists all supporting the same claim, I grew uncomfortable and started to withdraw from the community.

Concluding Thoughts

Right around the time that my role changed from that of an arbiter or honest broker to that of a critic, CCS entered a period of crisis: rising costs, lack of capital, declining attention to the climate problem, low CO_2 prices due to the financial crisis, and public resistance caused the increasing attention to stop. In hindsight, the criticism seems superfluous as CCS was already failing due to other reasons. However, I hope that it was not all for nothing and that the CCS community will be less likely to make the same mistakes as in the 2000s.

I remain convinced that CCS is a necessary evil and that it will at some point return to the limelight as one of the biggest potential contributors to global climate change mitigation. This may already be happening with biomass in combination with CCS (BECCS, which removes CO_2 from the atmosphere) to stay below the temperature limits in the Paris Agreement. When that happens, the CCS community might know better how to listen to its critics, engage sceptical audiences, and behave inclusively and responsibly. So far, there are few signs pointing in that direction. The arguments for BECCS, so far, sound very similar to those for CCS back in 2005. For sure, if the community again becomes a victim of groupthink, and a toy in the hands of much more powerful interests in the fossil fuel industry, the option is under greater threat of failing and climate change is even less likely to be prevented.

References

Bäckstrand, K., Meadowcroft, J., and Oppenheimer, M. (2011). The Politics and Policy of Carbon Capture and Storage: Framing an Emergent Technology. *Global Environmental Change*, 21(2), 275–281.

De Coninck, H. (2010). Advocacy for Carbon Capture and Storage Could Arouse Distrust. *Nature*, 463, 293.

De Coninck, H., and Bäckstrand, K. (2011). An International Relations Perspective on the Global Politics of Carbon Dioxide Capture and Storage. *Global Environmental Change*, 21(3), 368–378.

De Coninck, H. C., Flach, T., Curnow, P., et al. (2009). The Acceptability of CO_2 Capture and Storage (CCS) in Europe: An Assessment of the Key Determining Factors, Part 1 Scientific, Technical and Economic Dimensions. *International Journal on Greenhouse Gas Control*, 3, 333–343.

European Commission (2009). *Directive 2009/31/EC of the European Parliament and of the Council of 23 April 2009 on the Geological Storage of Carbon Dioxide and Amending Council Directive 85/337/EEC, European Parliament and Council Directives 2000/60/EC, 2001/80/EC, 2004/35/EC, 2006/12/EC, 2008/1/EC and Regulation (EC) No 1013/2006*. Brussels: European Commission, http://eur-lex.europa.eu/legal-content/EN/TXT/?uri=celex% 3A32009L0031.

Feenstra, C. F. J., Mikunda, T., and Brunsting, S. (2010). *What Happened in Barendrecht? Case Study on the Planned Onshore Carbon Dioxide Storage in Barendrecht, the Netherlands*, Energy Research Centre of the Netherlands (ECN) & Global CCS Institute (GCCSI). www.globalccsinstitute.com/publications/what-happened-barendrecht

IPCC (2005), *Special Report on Carbon Dioxide Capture and Storage*. Metz, B., O. Davidson, H. C. De Coninck, M. Loos, and L. A. Meyer (eds.) *Prepared by*

Working Group III of the Intergovernmental Panel on Climate Change. Cambridge: Cambridge University Press.

Van Egmond, S., and Hekkert, M. P., 2015. An Analysis of a Prominent Carbon Storage Project Failure: The Role of the National Government as Initiator and Decision Maker in the Barendrecht Case. *International Journal of Greenhouse Gas Control*, 34, 1–11.

10

Environmental Knowledge in Democracy

ESTHER TURNHOUT, WILLEM HALFFMAN,
AND WILLEMIJN TUINSTRA

10.1 Improving Environmental Knowledge

From the preceding chapters, it has become clear that the position of science is shifting. In the old conception, science was considered to be outside of policy and society, deriving its authority from this external position. More recently, strong entanglements between science, policy, and society are not only recognised, but also valued, and participatory forms of knowledge production are actively promoted in science policy. Yet, at the same time we also see a parallel trend in which the authority of science is increasingly being questioned. We only need to look at the science sections in the news media to recognise this. Just consider the multiple, and often contradictory, health effects attributed to the consumption of coffee and red wine, as well as the public controversies about climate change, the impact of neonicotinoids on wild and honeybee mortality, and the risks of genetically modified organisms. This situation is further exacerbated by recent developments in (populist) politics. The political campaigns around the Brexit referendum in 2016, in which a small majority voted in favour of the United Kingdom leaving the European Union, as well as the election of Donald Trump as president of the United States later that same year, have given rise to what many now understand as a 'post-truth' society. Such a post-truth society is characterised by a political and societal climate in which it has become increasingly common – acceptable, even – for public figures to skirt around the truth and to utter unchecked or false claims (also referred to as 'alternative facts').

In the context of these developments and trends, it is now more urgent than ever to reconsider the relation between science, policy, and society. This chapter discusses this relation from the perspective of democracy. Such a perspective highlights the importance of thinking about the position and role of science not just in instrumental terms – for example, as in improving

the uptake of knowledge or organising efficient science–policy–society interfaces – but also in terms of legitimacy and associated democratic values.
The chapter will discuss three such democratic values (accountability, diversity
and contestation, and humility) to serve as guidance for strengthening the
contribution of environmental knowledge to policy and society.

Although survey research shows that people's trust in science remains much
higher than for any other institution in society (e.g. De Jonge, 2016), the rise of
populist politics and its selective use of convenient scientific knowledge have
raised concerns among scientists and have resulted in repeated calls for
renewed respect for the authority and autonomy of science, in the media but
also in the different Marches for Science that were held around the world in
2017. However, these responses are largely inadequate (also see Sarewitz,
2016). While calls for greater autonomy of science to regulate itself and decide
on research priorities are certainly understandable in the current social and
political climate, it is important to recognise that scientists retreating back into
their ivory tower, avoiding engagement with policy and society, is unlikely to
restore respect for and trust in science (Brown, 2009). In the different chapters
of the book we have discussed some of the fundamental challenges for science
and its relation to policy and society. Specifically, we have discussed the
inevitability of value-laden judgements and uncertainties in the production of
knowledge, and we have pointed to the important role of societal, cultural, and
political context in how knowledge is both produced and used. We have
introduced the concept of framing to emphasise that environmental problems
do not simply 'exist' and that scientific facts cannot speak for themselves.
Instead, we – societal actors, policy makers, and scientists – construct stories
about issues and problems, including causes, solutions, and responsibilities,
and we bring certain scientific findings to bear on these issues and problem
frames. These basic insights all lead to the conclusion that science cannot be
seen as an unproblematic, independent, and detached way to observe and
describe objective reality, even if it sticks to its highest standards. Science is
not separate from, but a product of the reality it seeks to describe, and the
knowledge that science produces is necessarily partial, selective, and often
inconclusive, even if it often produces correct predictions and practical applications. And if science is itself the product of values, the expectation that it can
resolve conflicts of values and politics by itself is naïve and misguided, something that has become abundantly clear in the current, post-truth situation as
characterised above.

Drawing on the different chapters in the book, we can also identify a number
of ingredients for a more fruitful relation between science, policy, and society.
In addition to recognising the role of framing, values, and the limitations of

knowledge (Chapter 2, 3, 5), we have discussed the importance of building trust (Chapter 4), of open communication, clear responsibilities, and divisions of labour (Chapter 6), of non-elitist and symmetrical treatment of different kinds of knowledge and safeguarding diversity (Chapter 7 and 8), and of open reflection on the different expert roles that are effective and legitimate in different situations (Chapter 9). In the remainder of this chapter, we will use these ingredients to outline a vision of how environmental knowledge can play a legitimate and valuable role in democracy.

10.2 Beyond Technocracy

Earlier in the book (e.g. Chapters 6 and 9), we introduced the term 'technocratic' to refer to a specific way of framing environmental problems as technical: as problems that can be addressed by technological and science-based solutions. A domain of policy and governance where experts are in charge of defining problems and of designing, adopting, and implementing solutions is called technocratic. As we have discussed, technocratic forms of reasoning and decision making are prominent in the environmental domain, in the sense that it is usually scientific experts who identify and define environmental problems, attribute causes, and design and evaluate possible solutions. Of course, that does not mean that the will of experts automatically becomes rule. A strong role for expertise in environmental governance is often combined with an elected government or parliament that has the mandate and legitimacy to make decisions on the basis of considerations other than scientific expertise, or that can legitimately delegate decisions to experts.

Different theories and concepts of democracy have traditionally viewed the contribution of expertise to democracy in different ways (Brown, 2009). For some, any kind of indisputable authority – including science – is undesirable since this authority has not been established democratically, for example by means of elections. From this perspective, technocracy is seen as a direct threat to democracy. However, there are other views that are more positive about the role of science in democracy. From this perspective a prominent role of expertise in environmental governance can be justified by referring to the public good. Science is an authoritative representative of nature and environment – which are public goods – and science's presumed ability to provide indisputable facts is important for democracy because in doing so, it offers a common basis for political preferences to form and for political decision making to take place. In this sense, the task of the experts is to support the quality of democratic deliberation. In addition, technical reasoning can help achieve democratically decided plans and help governments to deliver what has been promised to the citizens. If democratic processes establish that certain levels of environmental protection are desirable, the expert could help to make it so. A limitation of such conceptions of experts' role in democracies is that they rely on sharp distinctions between expert responsibilities for 'the facts' and instrumental reasoning, while relegating values and goal identification to politics. This conception is ill suited for dealing with entanglement of facts and values, such as with the problems of framing, uncertainties, or knowledge controversies.

A third view is based on the recognition of both the importance of science for environmental governance and its inherent limitations. From this perspective, the contribution of environmental expertise can be improved by enabling forms of public participation in science. As discussed in Chapter 8, such participatory processes are assumed to make environmental expertise more robust and legitimate, resulting in better decisions and more effective implementation of those decisions. This participatory turn in science has run, at least partly, in parallel with wider trends in society towards participatory governance at multiple levels and scales. This also means that the primary audiences and users of environmental expertise have shifted from policy and government to nongovernmental and community organisations as well as citizens – hence our use of a science–policy–society interface instead of the more common, but also more limited, notion of the science–policy interface.

There are many initiatives for public participation in science and technology, including citizen science (see Chapter 8). These range from participatory biodiversity monitoring, identifying photos from Serengeti camera traps from your home computer (www.snapshotserengeti.org), public engagement processes organised for controversial technologies such as genetic modification (GM) or fracking (see Case C: Whose Deficit Anyway?), to the use of science to empower citizens and promote environmental justice (Ottinger, 2010). However, participation in science and technology does not always lead to desirable results. First, participation can easily fall into the trap of the information deficit model if the expectation is that participation will lead to a better understanding of and acceptance of science and technology (see Chapter 4). Second, participation is inevitably characterised by differences in power and authority which, if not managed carefully, can result in a dominant role of science and possibly even a silencing of other perspectives and knowledges (Pellizzoni, 2001; Turnhout et al., 2010). Moreover, many of these participatory processes are based on a deliberative ideal that holds that the open and rational exchange of views and arguments will lead to the best outcome. While good deliberative processes are valuable and important, their reliance on rationality is also risky. Specifically, deliberative and participatory processes can end up reproducing already dominant conceptions of what the best and most rational arguments are, thereby reinforcing exactly those technocratic tendencies that they were meant to overcome (Brown, 2009; Chilvers and Kearnes, 2015).

By pointing to these limitations, we do not suggest abandoning public engagement, deliberation, and participation and having scientists retreat back into their ivory tower. Rather, we suggest that science–policy–society relations need to be reconsidered in order to effectively move beyond technocracy and to ensure a legitimate role of science in democracy. This means that democratic norms will have to apply to science as well. The challenge for democracies is to identify principles and institutions that can accommodate the complications of blurred fact/value distinctions. In the next section, we will discuss three such norms that will contribute to the legitimacy and effectiveness of environmental knowledge at the science–policy–society interface.

10.3 Democratising Environmental Knowledge

10.3.1 Accountability

Accountability is an important concept in (representative) democracies. Citizens elect government officials, and these officials can be held accountable for their decisions and performance. The concept of accountability is also important in other domains. The public sector is accountable for how they spend public money, and companies are held accountable by the shareholders of the company. Accountability is not only about checks and controls; it is equally about the ability to give an account. This means that there is an expectation that representatives are able to explain and justify their choices and decisions.

This idea of accountability can also be applied to environmental expertise. In their knowledge-making practices, environmental experts create scientific representations of nature and the environment. There are two reasons why, from a democratic perspective, it is important that science is accountable for the way in which it represents the environment. The first is the political implications of environmental knowledge in framing environmental problems, shaping their solutions, and informing policy making and management. The second reason is the inevitable selectivity of scientific representations of the environment and the values and choices that inform this selectivity. Full representation is not only impossible, but also impractical. In his one-paragraph story 'On exactitude in science', Borges (1975) writes about a country where cartography had become so perfect that the map of the country was the size of a province. The purpose of the story is to illustrate the uselessness of representations that are too detailed and precise (also see Eco, 1994). In other words, representation requires simplification. However, as discussed in Chapter 3, this simplification is not neutral: it requires choices about what to include and what not, and these choices are guided by frames and values – those of the experts themselves or those of the expected users in policy and society – about what items of nature and environment are important. As in political representation, accountability in science requires connections between representatives and the constituencies that are represented by them (Turnhout, 2018). How to organise this accountability in practice can be a difficult question, since, at least in part, the constituencies of scientific representations are nature and environment themselves, and they cannot directly speak back to their representatives. Yet, they are affected by the way in which they are included or excluded from these representations, and by the way in which these representations are enacted in policy and management.

For the purpose of this book, we consider a legitimate role of science in democracy to be one in which science is required to give an account of the decisions, selections, and values that inform science, and in which those who have an interest in or are affected by environmental knowledge are able to hold science accountable for these decisions, selections, and values, and for the political implications of their knowledge (also see Turnhout, 2018). To organise this accountability, processes and institutions must be put in place that will involve some sort of participation or engagement of these actors with environmental knowledge-making so that the choices and values that inform this knowledge can be shared, discussed, and scrutinised. In addition, policy makers who make use of environmental knowledge have to be accountable for the way in which this knowledge has informed their decisions. The different examples and concepts of participatory knowledge production discussed in Chapter 8 can all potentially foster such accountability by creating connections between science, policy, and society. However, as discussed in the previous section, this has to be done in such a way as to avoid depolitisation and go beyond technocracy.

10.3.2 Diversity and Contestation

Democratisation involves opening up space for politics to take place (Mouffe, 2005). Without room for choice, deliberation, and contestation, there is no democracy. Applied to knowledge, democratisation is a response to the problems of technocratic forms of environmental governance and the depolitisation of the environment that is made possible by environmental expertise. In other words, environmental knowledge needs to become (re)politicised.

Such (re)politicisation of knowledge does not mean dismissing environmental knowledge by exposing how values and interests have informed it. Rather, it implies the recognition that nature and environment can be known and represented in diverse, but potentially equally legitimate, ways, none of which are neutral and value free. The diversity of scientific accounts arising from different disciplines or expert groups, each with their own particular frame, is not a threat to democracy, but a vital part of it.

This commitment to diversity is a way to avoid the polarisation that can often take place in knowledge controversies between different coalitions who all claim to speak *the* truth about nature and environment. We should not be under the illusion that there is only true knowledge that can be recognised unequivocally and distinguished from all other knowledge claims that then by default must be false. However, this does not automatically imply slipping into relativism where all knowledge is reduced to mere opinion and interest. Instead, this is about enabling diverse knowledge claims to be voiced and taken seriously. And this includes critical scrutiny and contestation of environmental knowledges: how they are made, how they frame environmental problems and solutions, and how they influence policy and management. Facilitating such processes requires avoiding the suppression of conflict and the premature closing of discussion as well as the distribution of power so that not just scientists are seen as authoritative knowledge holders (for an example, see Case I: The Loweswater Care Project).

Acknowledging diversity and allowing for contestation of knowledge are important for the accountability not just of science but also of politics. In other words, diversity and contestation are important for mutual relationships of accountability between science, politics, and society to flourish.

10.3.3 Humility

Many of the arguments developed in this chapter, and throughout the book more generally, come down to instilling an attitude of humility in science (Jasanoff, 2003, 2007). Accountability, diversity, and contestation require science to be open about its partiality and selectivity, and about the various limitations of environmental knowledge, while society needs to realise 'when to stop turning to science to solve problems' (Jasanoff, 2007, 33). It also requires science to be in open dialogue with other ways of knowing nature and environment and, instead of speaking truth to power, it requires science to speak the truth about its own power (Stirling, 2015) and to question those institutions and relations of power that prevent accountability, diversity, and contestation.

Humility is used here as the opposite of, and answer to, hubris. Hubris refers not necessarily to individual character traits of scientists, but to specific scientific practices or relationships where science is seen as the only way to get to the truth and as the only reliable source of knowledge for policy. We can also think of modes of knowledge production that are based on consensus or forms of knowledge integration that claim to offer completeness. These can be seen as hubristic to the extent to which they see alternative knowledge claims as irrational, false, or biased because they go against the consensus view or present a threat to the claim of completeness. In other words, while hubristic modes of knowledge production and use can be recognised by their closed nature and their appeals to truth and rationality, humility is about openness and diversity and welcomes the possibility to disagree.

Humility is not an easy virtue in the face of the constant requirement for scientists to show performance and relevance. In return for research funds or career opportunities, it can be tempting to over-promise, exaggerate certainty about environmental predictions, or downplay the validity of alternative accounts. It may even be tempting to soften doubt or uncertainties in order to guarantee the 'right' policy decisions. However, this is risky because, ultimately, these strategies backfire if opponents identify the gaps.

10.4 Democratic Expert Institutions

The responsibility for keeping expertise democratic does not rest on the shoulders of the individual expert alone. Research organisations and expert advisory organisations also have a responsibility for setting up accountability, for example by documentation of modelling practices, the origins of data sets, or the explanation of research priorities. Especially in research institutes that are required to generate their own income, accountability extends to openness about client relations, funding sources, and conflicts of interest. Through clear communication and stakeholder engagement, organisational accountability can extend beyond ritual 'transparency' that conceals actual accountability underneath an abundance of technical details.

Diversity and room for contestation are institutional responsibilities too. Public institutions should allow for variation in methods, perspectives, and approaches, rather than narrowing down the room for deliberation with technocratic means. This could mean that no discipline is allowed to monopolise police advice, or that alternative research approaches also receive public funding. The right to a second opinion or to appeal environmental assessments can be a procedural translation of such a principle. This may also mean that environmental expert organisations support some of their own opposition, or set up the fora to facilitate their own contestation. In confidently democratic expert institutions, diversity and disagreement are things to be celebrated rather than feared or silenced. Similarly, expert organisations have a responsibility for humility, for resisting the temptation to oversell their value or certainty, even when budgets are under pressure.

Ultimately, no hard rules and procedures can guarantee the spirit of democratic expertise. It is through practice and reflection that both individual experts and their organisations need to balance the tensions of democratic expertise: between the need to provide reliable facts and the need to show the limitations of these facts; between the self-confident trust in scientific standards and the humility required to be open about the way in which these standards frame environmental problems; or between the drive to deliver adequate environmental protection and the need to do so in fair deliberation of all concerns. Some democratic wisdom is required to reconcile the different functions of expertise in a democracy, i.e. to support democratic deliberation with both fact and reflection, to clarify the state of affairs in the world and explain the likely consequences of democratic choices, but also to help democratic governments deliver on their promises, and to question them when they are wrong.

References

Borges, J. L. (1975). Of Exactitude in Science. In *A Universal History of Infamy* (p. 131) London: Penguin Books.

Brown, M. (2009). *Science in Democracy: Expertise, Institutions, and Representation.* Cambridge: The MIT Press.

Chilvers, J., and Kearnes, M., eds. (2015). *Remaking Participation: Science, Environment and Emergent Publics.* New York: Routledge.

De Jonge, J. (2016). *Trust in Science in the Netherlands 2015.* The Hague: Rathenau Institute.

Eco, U. (1994). On the Impossibility of Drawing a Map of the Empire on a Scale of 1 to 1. In *How to Travel with a Salmon and Other Essays* (pp. 95–106). New York: Harcourt.

Jasanoff, S. (2003). Technologies of Humility: Citizen Participation in Governing Science. *Minerva*, 41, 223–244.

Jasanoff, S. (2007). Technologies of Humility. *Nature*, 450, 33.

Mouffe, C. (2005). *On the Political.* London: Routledge.

Ottinger, G. (2010). Buckets of Resistance: Standards and the Effectiveness of Citizen Science. *Science, Technology and Human Values*, 35, 244–270.

Pellizzoni, L. (2001). The Myth of the Best Argument: Power, Deliberation and Reason. *British Journal of Sociology*, 52(1), 59–86.

Sarewitz, D. (2016). Saving Science. The New Atlantis, Spring/Summer, 5–40. www
.thenewatlantis.com/publications/saving-science

Stirling, A. (2015). Power, Truth and Progress: Towards Knowledge Democracies in
Europe. In J. Wilsdon and R. Doubleday, eds., *Future Directions for Scientific Advice
in Europe* (pp. 135–153). Cambridge: Centre for Science and Policy.

Turnhout, E. (2018). The Politics of Environmental Knowledge. Accepted for publication by *Conservation and Society*.

Turnhout, E., van Bommel, S., and Aarts, N., (2010). How Participation Creates
Citizens: Participatory Governance as Performative Practice. *Ecology and Society*,
15(4), 26.

11

Conclusion

Science, Reason, and the Environment

WILLEM HALFFMAN, WILLEMIJN TUINSTRA, AND ESTHER
TURNHOUT

In this book, we have identified and analysed key hurdles that environmental scientists encounter when providing knowledge for collective decision making. One crucial obstacle is the rigidity of problem frames. Scientists and policy makers alike have a tendency to frame problems as well defined and well structured, because this makes it easier to develop straightforward solutions. Natural scientists[1] in particular are trained to first pin down a problem with a fixed definition, before doing research. Well-defined problems have clear demarcations, relatively unambiguous and uncontested definitions of terms, and a clear set of values and objectives at stake. Under such conditions, instrumental reasoning comes into its own, with precise measurement, established methods, and formalised procedures to identify optimal solutions. This is where the 'hard' sciences can fully deploy their unquestioningly formidable potential.

In contrast, environmental problems are often *ill* defined, unstable, contentious, ambivalent, political, messy, and fraught with uncertainties. When faced with these contestations, and the diverse perspectives and values involved, the standard instrumentalist approach runs into limitations. Throwing hard facts at policy to make it do things only works when decision makers are looking for exactly *these* facts, at *that* very moment, and only if they recognise the source as legitimate. Such a match is not impossible, but it is typically established after long collaboration, creating mutual agreement on what is credible, legitimate, and salient knowledge (Bijker, et al., 2009; Cash, et al., 2006; De Vries, et al., 2010). But even if such a match between the needs of scientists and policy makers might result in effective and smooth interactions,

---

[1] In case you skipped directly to this concluding chapter, and just to make sure, we stress again the use of 'science' in its continental European connotation, to include all academic disciplines. We use 'natural' or 'social' science when specificity is needed.

this does not mean that this role of expertise is necessarily legitimate, since such close collaborations can also be elitist and exclusionary (see Chapter 9).

Hence, we explained in Chapter 2 why the 'trust me, I'm a scientist' approach is only productive and legitimate under very specific circumstances. Referring to academic training, 'the scientific method', or the institutions of science in general (such as peer review) may provide good arguments to call some knowledge more reliable than others. However, it does not universally distinguish hard truth from nonsense, and rarely succeeds to automatically convince an increasingly educated public. After all, as we showed, science itself is a diverse landscape of knowledges, with different styles and specialties relying on varying procedures and principles to certify knowledge claims. Hence, some use the plural 'sciences' to denote this family of knowledge practices, rather than one homogeneous 'Science'.

Acknowledging knowledge diversity clearly complicates things, but it also prepares for a more meaningful cooperation across boundaries. The alternative is no longer an option. Imposing instrumental reason, either by technocratic experts or policy makers, tends to deny the societal diversity in environmental concerns, understandings, valuations, and hopes. The attempt to 'rationalise policy' in this way often enforces the problem definitions of the powerful and presents biased technical reasoning as neutral and universal (Flyvbjerg, 1998; Scott, 1998; Wynne, 1982), excluding other potentially relevant perspectives and knowledges. Ultimately, such a strategy will also backfire, such as when dissenting voices refuse to remain silent, or when solutions turn out to be ineffective because they fail to address the full complexity of the problem.

So, what to do? How to deploy the considerable benefits of the sciences, while also recognising these complexities? As a first step towards a more productive interaction, we presented frame analysis in Chapter 3. Frame analysis helps to identify differences in how environmental problems are understood, and how facts are perceived as relevant for the problem at hand. The chapter offered conceptual tools to identify frames in language, but also frames embedded in images, procedures, or even devices. This prepares the way for frame *reflection* in practice and potentially deeper forms of social learning among those involved in dealing with environmental problems. Rather than arbitrarily fixing problem definitions, frame reflection can open them up for reasoned debate or collective deliberation.

Diverging frames are one of the reasons for the failure of the 'linear model': the one-way conveyance of factual scientific knowledge to policy. The idea that science can simply deliver value-free facts and inform decisions in a neutral way is an extensively criticised, but rather stubborn misconception. This also applies to the information deficit model: the idea that methodical, certified, clearly explained research should instantly convince participants in public decision making. The complications of environmental controversies (and especially heated, mediated ones) show the failure of the linear model most clearly: actors in controversies typically question what may count as reliable knowledge, who can be trusted to provide knowledge, how to assess uncertainties, what is part of the problem and what not, and how causal connections lead to responsibilities to act, as explained in Chapter 4. Both cases in this chapter (fracking and climate change) demonstrate how controversies revolve around trust in the institutions that produce expertise and not just the correctness of facts. In times of controversy, building trust may therefore be a priority over being right; a process that typically requires patient social skills from all involved, in addition to expert knowledge of methods or statistics.

In Chapter 5, we inspected more closely the elements of uncertainty in environmental knowledge. Here too, more is at stake than just the precision of instruments or the adequacy of data sets – the technical factors in uncertainty. Inherent variability of the environment (especially the social environment), and diversity in populations or in perspectives on what is to be protected, may create unstable conditions for technical reason. This has profound consequences, for example for risk analysis and the policies based on it, as risk analysis tends to downplay its uncertain fringes and is ill-equipped to address them. Defining the conditions for what is to be counted in risk calculation, and how to measure and count it, requires more than a technical committee of experts and privileged stakeholders to establish standards (Chapman, 2007). A deeper acknowledgement of uncertainties also results in a more sophisticated understanding of variation between types of risk, and, consequently, more developed risk management strategies (Renn and Klinke, 2004).

Chapter 6 turned to the issue of usable knowledge and explored when knowledge is considered usable and what conditions have to be met for its production. Various scholars have developed lists of key requirements for usable knowledge. However, making knowledge usable is not just about the features of the knowledge itself, but also about understanding the setting in which it is to be used. How controversial is the issue? What are diverging concerns? What is the phase of the policy process? What is the dominant decision-making style in the policy field? Knowledge usability also requires an understanding of the precise expert role. After all, experts do far more to support public decisions than just provide certified facts: their role might include devising strategies, mediating in conflicts, or devising practical solutions.

Over time, such interactions develop into routines: they turn into procedures and become part of organisations. Boundary organisations or boundary objects may form, which structure the cooperation between experts and decision makers. However, the mere fact that such structured cooperation exists is no guarantee for fair or effective policy, as the fisheries case in this chapter showed. Institutionalised expert/policy cooperation may result in a well-oiled advisory 'machine', but not necessarily in legitimate or sustainable outcomes. For example, knowledge diversity may be 'amputated' by such institutions for the sake of simpler procedures, as in fisheries advice that relies on quite particular forms of population modelling.

Hence, Chapter 7 asked the question of how to bring diverse knowledge together and prevent such 'amputation'. One prominent approach in the environmental sciences has been to 'integrate' knowledge – that is: to make different forms of expertise commensurate, such as in the context of an encompassing model of economic–environmental interactions. This is a strategy of combining interdisciplinary knowledge that has been used to great effect, but one that also has some disadvantages. For example, conceptions of how people act, decide, or behave vary drastically between scientific disciplines, especially in the social sciences. Hence, 'integration' may force incompatible understandings of people or of the environment into the straightjacket of a computer model, reducing deeper conceptual knowledge diversity for the sake of compatible, exchangeable numbers. However, this is not the only possible approach to knowledge integration: there are also forms of interdisciplinary cooperation that leave more room for conceptual differences. The cases in this chapter show that not all knowledge integration has to result in a cognitive straightjacket.

The challenges of combining heterogeneous forms of knowledge seem particularly sticky in the case of lay expertise: the knowledge held by people who are not professional or trained scientists. While lay expertise and science are different, in practice this distinction is often more blurred than such binary terms suggest. Chapter 8 investigated the value of lay expertise and explored the surprisingly dynamic forms of cooperation between lay or volunteer 'researchers' and professionals, especially in biodiversity research. In the context of specific practices of conservation or natural history, cooperation has been carefully fostered and has grown into immensely valuable knowledge resources. This is particularly well illustrated in the cases at the end of the chapter, including a wealth of examples from botanical gardens all around the world. The predominant approach of these cooperation projects has not been to put one community in service to the needs and concepts of the other, such as when volunteers become data-collecting drones for researchers, but one of carefully fostered mutual understanding that respects the diversity of actors, perspectives, and knowledges.

Chapter 9 summarised the work of experts providing knowledge for public decision in three different modalities or overall strategies to position oneself as an expert: that of a servicing expert, an advocating expert, or a diversifying expert. The technical provider of solid knowledge for well-structured problems from this chapter's opening now returns as one particular strategy, which may indeed be viable and meaningful under specific conditions. However, experts may also feel the obligation to speak up and take a more advocating role, as the carbon capture and storage case illustrates. This does not mean such experts throw science out the door and turn to blatant politicking. On the contrary, they may speak up because they no longer think the fundamental assumptions of the servicing experts are scientifically tenable. In turn, diversifying experts may actually look for alternative conceptions, or attempt to mediate between forms of knowledge. At the end of our analysis, the range of options for experts has expanded far beyond a role that consists of just providing solid facts.

So, how do you choose? Or, on a collective level, how do we decide what kinds of expert organisations and institutions are needed in democratic societies? In the end, far more is at stake than just the methodological soundness of knowledge. Other values are relevant here – values that may not sound very 'scientific', but that are nevertheless crucial to how we organise environmental knowledge and decision making justly and democratically. They include accountability, respect for diversity of values and knowledge, the right to contest expert claims, and, ultimately, an appropriate humility of expertise. The question is not just how to organise expertise to be right, but also how it should function in an open, democratic society. Ultimately, the precise shape of these institutions should be the outcome of democratic deliberation (and certainly not dictated by scientists and/or experts), but we hope we have at least laid out some of the terms in which this deliberation can be cast. We have aimed to provide a repertoire for reasoning about scientific expertise that is more sensitive to democratic values, to knowledge diversity, and to the limitations of technocratic approaches and instrumental reason.

We recognise that with this appeal for diversity, democratic values, and a more sensitive and humble expertise, we go against current tendencies in the opposite direction. In the face of callously populist leaders, ruthless corporate buccaneers, or scheming self-interested lobbies, defenders of technocratic expertise may argue that they guard against such brutality with a firewall of disinterested science. They may argue that our messages weaken their argument for rational and strong environmental protection policy and management. However, throughout the book, we have shown how power already operates within expertise, not just in its use, by fixing problem definitions, or altering what counts as acceptable evidence. Too often, technical reason has been abused as an excuse to protect vested interests: to claim that organochlorine pesticides were a sensible instrument for a more rational agriculture, that there was no solid scientific evidence for climate change, that biofuels would offer a sustainable solution, or that nuclear accidents were extremely improbable. In our view, more diversity, accountability, and humility in expertise actually contributes to a more equitable distribution of power. In that respect, the argument developed in this book can be seen as one of *advocating* the importance of *diversifying* expertise.

References

Bijker, W. E., Bal, R., and Hendriks, R. (2009). *The Paradox of Scientific Authority: The Role of Scientific Advice in Democracies*. Cambridge: MIT Press.

Cash, D. W., Borck, J. C., and Patt, A. G. (2006). Countering the Loading-Dock Approach to Linking Science and Decision Making: Comparative Analysis of El Niño/Southern Oscillation (ENSO) Forecasting Systems. *Science, Technology, and Human Values*, 31(4), 465–494.

Chapman, A. (2007). *Democratizing Technology: Risk, Responsibility and the Regulation of Chemicals*. London: Earthscan.

De Vries, A., Halffman, W., and Hoppe, R. (2010). Policy Workers Tinkering with Uncertainty: Dutch Econometric Policy Advice in Action. In H. K. Colebatch, R. Hoppe, and M. Noordegraaf, eds., *Working for Policy* (pp.91–109). Amsterdam: Amsterdam University Press.

Flyvbjerg, B. (1998). *Rationality and Power: Democracy and Practice*. Chicago: University of Chicago Press.

Renn, O., and Klinke, A. (2004). Systemic Risks: A New Challenge for Risk Management. *EMBO reports*, 5(S1), S41–S46. doi:10.1038/sj.embor.7400227

Scott, J. C. (1998). *Seeing Like a State: How Certain Schemes to Improve the Human Condition Have Failed*. New Haven: Yale University Press.

Wynne, B. (1982). *Rationality and Ritual: The Windscale Inquiry and Nuclear Decision Making in Britain*. Chalfont St Giles: BSHS.

Index

Printed in the United States
by Baker & Taylor Publisher Services